Validation of Food Preservation Processes Based on Novel Technologies

Validation of Food Preservation Processes Based on Novel Technologies

Edited by

TATIANA KOUTCHMA
Guelph Research and Development Center, Agriculture and Agri-Food Canada, ON, Canada

ELSEVIER

ACADEMIC PRESS
An imprint of Elsevier

Academic Press is an imprint of Elsevier
125 London Wall, London EC2Y 5AS, United Kingdom
525 B Street, Suite 1650, San Diego, CA 92101, United States
50 Hampshire Street, 5th Floor, Cambridge, MA 02139, United States
The Boulevard, Langford Lane, Kidlington, Oxford OX5 1GB, United Kingdom

Notices

Knowledge and best practice in this field are constantly changing. As new research and experience broaden our understanding, changes in research methods, professional practices, or medical treatment may become necessary.

Practitioners and researchers must always rely on their own experience and knowledge in evaluating and using any information, methods, compounds, or experiments described herein. In using such information or methods they should be mindful of their own safety and the safety of others, including parties for whom they have a professional responsibility.

To the fullest extent of the law, neither the Publisher nor the authors, contributors, or editors, assume any liability for any injury and/or damage to persons or property as a matter of products liability, negligence or otherwise, or from any use or operation of any methods, products, instructions, or ideas contained in the material herein.

British Library Cataloguing-in-Publication Data
A catalogue record for this book is available from the British Library

Library of Congress Cataloging-in-Publication Data
A catalog record for this book is available from the Library of Congress

ISBN: 978-0-12-815888-3

Publisher: Charlotte Cockle
Acquisitions Editor: Nina Rosa de Araujo Bandeira
Editorial Project Manager: Rachel Pomery
Production Project Manager: R. Vijay Bharath
Cover Designer: Victoria Pearson

Typeset by MPS Limited, Chennai, India

Dedication

In the memory of my mother
Nina M. Kuchma
who died in 2020.

Contents

List of contributors

Larry Keener
International Product Safety Consultants, Seattle, WA, United States

Tatiana Koutchma
Novel Food Processing Technologies, Agriculture and Agri-Food Canada, Guelph, ON, Canada

Keith Warriner
Food Science Department, University of Guelph, Guelph, ON, Canada; Center for Public Health and Zoonoses Canada

Preface

With the fast-growing demands in mild heat and cold food preservation and sanitation markets, novel processing technologies have been emerging in food production for extending products' shelf-life and eliminating foodborne pathogens. A science-based standardized validation process is a critical component in assessing the efficacy of novel technologies as a microbial control measure according to food regulations and successful commercialization. While the longer history of validation of thermal technologies resulted in advancing standard methodologies, the approaches for the validation of nonthermal technologies are not established and are currently under development. Among a number of technologies that are presently investigated for premium food processing, high hydrostatic pressure (HHP) and ultraviolet (UV-C) light are nonthermal alternatives to pasteurization approved by international regulatory agencies and recognized as the simplest and environmentally friendly ways to destroy pathogenic and spoilage organisms with lower impact on quality and nutrition values and sensory attributes of many products.

Despite numerous publications that are available on scientific, technological, and engineering aspects of nonthermal processing applications for food treatment, no recent monograph exists that would integrate fundamental knowledge about principles and scientific approaches of food process validation on all stages of process development and commercialization. The need for clear knowledge and understanding of validation steps of nonthermal processes is vital. This book is a first attempt to summarize existing scientific experience of validation of traditional thermal and nonthermal technologies, such as canning or radiation, and recommend key activities for bench top, pilot, prototype, and commercial validation of HPP and UV systems for the treatment of variety of food products. The objective and focus of this manuscript is to review definitions, principles, and practical approaches of food process validation and summarize the experience of thermal processing and of a few commercially available novel processes that have been using HPP technology, recently ultraviolet light and other novel technologies for safety and preservation operations. The aspects explored include available process applications for liquid and solid foods, critical product and process control parameters for each technology, microbial challenge studies including selection and recommendation of surrogates, and consideration for procedures to measure and control the delivery of the required novel process, quality, and equipment validation. In addition, sanitary design and equipment scalability will be discussed to facilitate the implementation of nonthermal technologies and mitigate risks during food production.

The introduction chapter starts with reexamining the history of thermal process validation and verification and existing practical methods and approaches of thermal preservation process establishment and validation. Also, new challenges of development, validation, and commercial implementation of novel nonthermal processes compared to thermal processing are outlined including production benefits and review of fundamental principles.

In Chapter 1, General Principles and Approaches of Food Process Validation, general concept and definitions, structure, key components, and approaches to process validation are presented and discussed in detail. This includes physical, microbiological safety, quality, and equipment validation with elements of cleaning, calibration and analytical parts, and validation of facilities. The general guidelines for each key component of the validation process are given along with the basic outline of objectives and critical procedures of process and equipment validation. Essential role and requirements for process authority are presented in this chapter.

The focus of Chapter 2, Validation of High Hydrostatic Pressure Process, is the validation of HHP process of HPP that is used for nonthermal pasteurization. The fundamental principles of HPP along with essential practical considerations in terms of critical product and process parameters, packaging, microbial challenge studies, and other testing and scaling up issues are discussed in details to assist HPP end users to accelerate process innovation.

The goal of Chapter 3, Case Study of Validation of High Hydrostatic Pressure Processing of Fruit and Vegetable Smoothies in PET Bottles. Chapter 4 is to discuss Validation of Light-Based Processes, such as continuous UV-C light, pulsed light, and LEDs, for the treatment of surfaces and liquid products and provide guidelines for technology implementation. Information regarding international regulatory requirements on light technologies is summarized and will be helpful in determining validation objectives and scope. Chapter 5, Case Study of Validation of Ultraviolet Light Pasteurization of Sugar Syrups, presents the methodology and validation of microbiological efficacy of commercial UV treatment of sugar syrups.

Chapter 6, Overview of Other Novel Processes and Validation Approaches, is focused on the overview and brief discussion of principles, commercial applications, and main validation approaches of two commercially available group of technologies, such as ionizing irradiation and advanced microwave heating and emerging treatment using cold atmospheric plasma. Pros and cons of each technology will be presented along with status update on regulatory approvals and validation procedures. Due to the physical nature of those technologies, the approaches for process validation differ and face different technical challenges. For this reason, different approaches can be used to demonstrate and document the targeted performance.

Not all nonthermal technologies and validation issues were discussed in this book. In general, the book content is aimed to assist and provide recommendations for food processors, technology developers, equipment manufacturers, regulatory inspectors, and extension specialists during the commercialization of novel technologies. Moreover, the suggestions for validation protocols and reports can be helpful for intended audience. However, this book represents the most comprehensive and ambitious step on the subject of validation of novel technologies for foods that exists to date. Also, the important step forward has been made in scientific understanding of the needs and practical tasks for the future.

Introduction

Tatiana Koutchma
Novel Food Processing Technologies, Agriculture and Agri-Food Canada, Guelph, ON, Canada

1. History of thermal process validation and verification

1.1 Approaches to thermal process validation

Throughout the history of food processing, the application of preservation technologies has been traditionally associated with four fundamental concepts:

1. Application of thermal energy to elevate product temperatures to achieve long-term or extended safety, stability, or preservation;
2. Removal of thermal energy to reduce product temperature and extend shelf-life;
3. Removal of water from product structure or to acidify foods and thus achieve an extended shelf-life;
4. Application of packaging or a step required for maintaining of product properties achieved after processing during storage.

The application of thermal energy to elevate the product temperature has been one of the key processes in the food-processing industry, which often aims to produce commercially sterile food products and pasteurized products of optimal quality.

Depending on the product type—solid or liquid—being processed and the shelf-life required, thermal preservation can be accomplished in two ways. In the case of solid foods, heat is applied to a filled, sealed container, and in the case of liquids, the sterilization or pasteurization of products can take place outside of their final container and before filling in a clean or aseptic environment.

Sterilization and pasteurization operations rely, respectively, on elimination and reduction of the most resistant pathogens of public health significance and spoilage microflora. Commercial sterility is defined as: the condition achieved by application of heat which renders food free from viable microorganisms, including those of known public health significance, capable of growing in the food at the temperatures at which the food is likely to be held during distribution and storage (US Department of Health, 1994). The term "pasteurization" was originally defined as a process of mild heat treatment to reduce significantly or kill the number of pathogenic and spoilage microorganisms. The definition of a traditional pasteurization process relied only on thermal treatment and is achieved by exposing foods to heat for a certain amount of time. Unlike sterilization, after pasteurization the food is not free of microorganisms since heat treatment is not severe enough to kill heat-resistant spores that can survive the process and may be present. Therefore additional forms of preservation such as refrigeration (e.g., milk), atmosphere modification (e.g., vacuum packaging of meats and cheeses), addition of antimicrobial preservatives (e.g., sodium chloride, sodium nitrite, ascorbic acid, sorbic acid, and sulfur dioxide), or combinations of these techniques, are required for product stabilization during distribution.

Since introducing preservation principles in food processing at the start of the 19th century, the food industry and scientific community have accumulated critical knowledge for the establishment and validation of thermal and other traditional processes that includes:

1. Established organism of public health concern for different groups of products such as low-acid products, milk, and juices;
2. Well understood and known kinetics of microbial destruction and its parameters of thermal resistance (z), decimal reduction time (D_T), and lethality;
3. Knowledge of product heating and heat transfer in given processing systems;
4. Establishing of equivalent safety of different processing systems expressed in "lethality" terms.

Process validation is necessary to give an assurance that commercial sterility, pasteurization, or other related food safety objectives (FSOs) that have been achieved in test conditions also will be delivered in all subsequent production conditions. Historically, there are two approaches to the validation of thermal processes.

1. The first approach has been to use direct microbial kill measurements involving inoculation studies with spore or bacterial cell suspensions with known heat destruction/resistance characteristics or other microbial challenge tests.
2. The second approach involves the collection and recording of time and temperature data history in the tested product using specialized data-logging systems, and more recently in-process data-logging with small data loggers that can pass through the continuous process, and collect data for subsequent analysis and interpretation.

In both cases some form of mathematical modeling has been used to interpret the data in the context of FSOs and/or commercial sterility. The mathematical models for interpretation of the data are designed to reflect the appropriate death kinetics of the target marker organisms. Mathematical models such as the application of D and z values in the development of thermal processing have proven extremely useful. The models also provide a way of integrating the overall lethal effect of the thermal process, enabling processes of different times and temperature history to be compared on a common basis of lethality equivalence. Lethality is an essential element of process establishment and validation that has been developed for thermally processed foods. It provides the ability to express process delivery in terms of units of lethality and to understand the equivalent safety of different processing systems. By incorporating both the kinetics of the destruction of the organism of public health concern and the mathematical modeling of heating curves the thermal processing industry developed the terms of process values such as "sterilization value" (F_T) and pasteurization value (P_T). When the sterilization value (F_T) has a reference temperature of 121.1°C it is called an F_o, and pasteurization value (P_T) with a reference temperature of 72°C is called P_o. The sterilization and pasteurization values are the measurement of the number of minutes that would result in the same effectiveness of the process if the product had been held at 121.1°C or 72°C/90°C, respectively. A "worst case" approach by choosing samples that received the lowest time–temperature treatment to establish a safe process is traditionally used in thermal processing. This approach aiming at giving a greater assurance of process safety often results in a vast majority of overprocessed products and leads to lower quality as a result.

The ability to understand the equivalent efficacy of thermal processes is important when it comes to commercial applications and compliance with associated international regulatory requirements. Also, it is evident that the validation template developed over many years for activities such as canning, and aseptic applications could serve as a "blueprint" for other applications of novel processing technologies.

1.2 Establishment of the preservation process

The critical product, process, and storage condition differences between target microorganisms that are to be considered to establish pasteurization or sterilization preservation specifications are summarized in Table 1.1.

Table 1.1 Critical product, process, and storage conditions for the establishment of preservation specifications.

	Pasteurization			Sterilization
Product parameters				
pH	<3.5	3.5 <pH < 4.6	pH > 4.6	pH > 4.6
A_w				>0.86
Process parameters				
Temperature, °C	65–72	>65	>65	>121
Additional hurdles	No	Refrigeration	Antimicrobials, A_w	
Target microorganisms				
Pathogenic	*E. coli* *Listeria* *Salmonella*	*E. coli* *Listeria Salmonella*	Nonproteolitic *C. botulinum*	*C. botulinum* spores
Spoilage	Molds, yeasts	Lactic bacteria, yeasts, molds		*Geobacillus* spp. *Bacillus cereus*
Storage	Ambient	Refrigerated conditions		Ambient temperature
Packaging		Hermetically sealed flexible containers		Hermetically sealed flexible containers

A general approach to the establishment of processing technologies includes:

1. Identification of the organism of concern of public health significance;
2. Identification and selection of the appropriate target end-point or microbial reduction;
3. Development of a conservative estimation of the ability of the process to consistently deliver the target end-point or specific log reduction;
4. Quantitative validation (microbiological or mathematically) of the lethal treatment delivered;
5. List of the critical factors and procedures used to control the delivery of the required process.

For the design of a "preservation specification" to meet required specific logarithmic microbial reduction (SLR) in numbers of target microorganisms, processing time F_p or target process value is traditionally defined by the initial load of target organisms (N_o), the end-point of the process (N_F), and the logarithmic resistance of target bacteria under defined conditions (D_T) as given in Eq. (1).

$$F_p = D_T(\log N_o - \log N_F) = (\text{SLR})\, x D_T \tag{1}$$

where D_T is a decimal reduction time (min) at the reference temperature of a thermal process.

Databases containing kinetic inactivation parameters for various target pathogenic and spoilage microorganisms are needed for the establishment of processing time and critical preservation process parameters.

Eq. (2) estimates the accumulated lethality F, sterilization (F_0), or pasteurization value (P_0) of a thermal process, when the reference temperature and z values are defined for specific microorganisms and products

$$F_o = \int_0^{\text{time}} 10 \frac{T(t) - T\text{ref}}{Z} \text{dt} \tag{2}$$

where T (°C) is a temperature and t is time (min). T_{ref} is a reference temperature of the thermal process. Z is bacterial cells or the spore's thermal resistance (°C) at the reference temperature.

The validation of the processing time under actual production conditions requires the accumulation of a minimum process value under worst case conditions (cold-spot, low process temperature, etc.) of F_o. The measured or actual treatment time F_p under similar worst case conditions that is required to achieve a target SLR should meet the condition (3) that is:

$$F_o > F_p \tag{3}$$

The procedure of establishing a process and then validating its effectiveness is dependent on the processing system used and the food product being produced. However, all assumptions associated with historical experimental procedures will need to be verified for their applicability to the new application. The processor needs to make sure that a sufficiently verified approach is used to collect the needed information necessary to understand the process delivered and if the target end-point was achieved.

The conclusions and recommendations that exist for the establishment and validation of thermal pasteurization and sterilization treatment can be modified and adopted for novel thermal or nonthermal processes. It will be necessary to determine microbial inactivation kinetics and rate constants that are similar to *D*- and *z* values and integrated process lethality to determine process equivalency.

2. Nonthermal technologies and challenges of validation of novel processes compared to thermal processing

2.1 Review of fundamental principles

Over the last few decades, the fifth "alternative" processing concept called "novel processing technologies" started to emerge globally in food production. In the history of technology development, novel or emerging technologies are those contemporary technical innovations which represent progressive developments within a field for a competitive advantage. In the food industry, as a result of modern demands for foods that are fresher, healthier, more natural, or minimally processed and additive-free, novel processing techniques are currently under broad development and implementation. The use of novel technologies ultimately brings the advantage of the production of foods with better quality and higher nutrient contents due to reduced thermal and chemical abuse, superior safety attributes during extended shelf-life, and at a reasonable cost for the consumer. Food safety is one of the essential components that drives the development of novel technologies such as physical microbial intervention to reduce, control, or eliminate foodborne pathogens from food products and contact surfaces. Additionally, novel processing technologies have potential to be used as tools to tailor foods with better quality, added or enhanced functional and nutritional values, to lower the carbon footprint, and substantially reduce water volumes used in heat transfer processes. As an example, novel processing methods are being explored to potentially create hypoallergenic products, and have been tested to alter the allergen reactivity of components of food matrices.

This fifth "novel" processing concept includes advanced thermal and nonthermal high-technology methods based on the utilization of mechanical, electrical, and electromagnetic energy, and combined applications approaches. So-called novel foods with no history of consumption that may be produced by novel processes have started to emerge in the market and often can replace traditional style foods. For example, old-style ham treated by high hydrostatic pressure (HHP) or fresh apple cider pasteurized using ultraviolet (UV) light for better product safety are described as novel foods.

Food-processing techniques that can be used for the reduction of microbial loads include removal and inactivation approaches. Known removal unit operations include filtration, centrifugation, and separation, which belong to the group of hydrodynamic processes that are driven by a hydrostatic or hydrodynamic pressure gradient. Unit operations that are using microbial inactivation include a number of physical methods and chemical agents capable of pronounced bactericidal and/or sporicidal effects. A summary of different techniques that have been explored and their fundamental principles to deliver a lethal microbial treatment is presented in Table 1.2.

There are typically numerous unknowns associated with the safe use of novel preservation technologies in the production of foods and beverages intended for human consumption. Due to different fundamental principles and performance capabilities of novel technologies, novel processes differ from traditional thermal processing in terms of the types of food categories that can be treated, microbial efficacy and microbial destruction modes, desired and undesired effects on food composition and quality, and their economic and environmental impacts. Additionally, the full commercialization and production potential of novel thermal and nonthermal technologies and food-processing operations based on their use is not being realized due to a lack of informed and endorsed guidance regarding how to implement the proper technology validation and demonstrate its safety. Before a novel process can be used and the resulting product can be sold, thorough reviews of the impact on composition, quality, and nutrition, and evaluations of the safety of the novel process and consequently novel foods have to be conducted by food industry professionals and regulatory agencies. As a result, the incremental developments of novel technologies for a variety of food applications have led to

Table 1.2 Food-processing unit operations and fundamental principles of traditional and novel processing technologies.

Microbial reduction principle	Physical mechanism of action	Driving force	Process	Additional bactericidal modes
Removal	Momentum transfer	Pressure difference	Filtration Separation Centrifugation	
Thermal inactivation	Heat transfer	Temperature difference	Blanching Drying Heat pasteurization Heat sterilization	
	Electromagnetic energy	Temperature difference	Infrared heating Microwave heating (MWH) Radiofrequency	
	Electrical energy	Temperature difference	Ohmic heating Conduction heating	
	Momentum transfer and heat generation	Temperature difference	Pressure-assisted thermal sterilization	Pressure and compression heating
	Momentum transfer and heat generation	Temperature	Ultrasonication	Cavitation Local temperature increase
Nonthermal inactivation	Electromagnetic energy	Charged particles Electrons transfer UV light photon transfer	Irradiation Electron beams UV energy	Free radicals
	Supercritical fluid conditions	Concentration	Pressure and CO_2	High penetrating power
	Momentum transfer	Hydrodynamic pressure in liquids Temperature increase	Pressure homogenization	Temperature Cavitation
	Momentum transfer	Hydrodynamic pressure waves in solids	High hydrodynamic pressure/ shock waves	Shear stress
	Momentum transfer	Hydrostatic pressure in liquids and solids	HHP	

(Continued)

Table 1.2 (Continued)

Microbial reduction principle	Physical mechanism of action	Driving force	Process	Additional bactericidal modes
Pulsed energy	Electrical fields	High voltage gradient	Pulsed electric fields (PEFs)	Temperature
	Electromagnetic energy	Light and UV photons transfer	Pulsed light	Temperature
	Momentum transfer	Pressure	Pressure cycling	
Plasma	Electrical or electromagnetic energy	Charged particles and photon transfer	Atmospheric plasma	Light photons Free radicals Charged particles
Chemical	Oxidation	Concentration	Ozone Carbon dioxide Ion exchange Chlorination pH adjustment	
	Osmotic pressure	Solute concentration	Water activity	Bound water

significant accumulation of new scientific and engineering information. However, the technological readiness level is assessed by its suitability for the actual application in its final form.

2.2 Challenges and perspectives of commercial implementation

Paramount among these are concerns for food safety, validation, and also regulatory acceptance. Confirming food safety is a daunting undertaking because validating safety of food technology involves every element of the manufacturing process. Consideration must be given to microbial stability, toxicology, product and packaging interactions, process capability, chemical stability, and also possible harmful effects of the stabilization process. In most cases there will also be the need to provide objective data that confirm, with a high degree of confidence, that the product meets the safety expectations of the various regulatory agencies. Under certain circumstances the rigor demanded for confirming safety and regulatory compliance, when joined with the high associated cost, has served to slow or deter the commercialization process and therefore also the uptake of novel processing and preservation methods.

The study of the commercialization of novel technologies and their applications around the world (Jermann et al., 2015) has been based on the results of two surveys in North America and the European Union. The respondents to the surveys were asked to identify novel technologies either applied now or with the potential to be commercialized in 5–10 years, commercialization factors, associated regulations, and limitations. In the North American survey, high hydrostatic pressure processing (HPP), MWH, and UV light were the main technologies applied now and anticipated to be commercialized in the next 5 years. PEFs technology was placed third in preference to UV in the EU survey; HPP and microwaves were identified as the main technologies now and in the next 10 years. Also, there were geographical differences, with North America finding UV light and radiation, and Europe finding PEF of more importance currently. Cold plasma and PEF were anticipated to be more important in Europe in 10 years' time while HPP, MWH, and UV remained more important in North America.

Higher quality products, product safety, and shelf-life extension have been identified as the main applications for novel technology development. When answering questions about the main drivers of innovations, equipment manufacturers, academia government research organizations, and large food corporations were selected as the three key players who contribute to innovations in the food industry.

The implementation and potential for implementation of novel technologies were also assessed for different industries, including meat and poultry, dairy, fresh produce (including fruits and vegetables), drinks and beverages, and fish and sea products.

According to the surveys, this is how these technologies scored (in decreasing order):

1. Meat and poultry sector: *HPP > MWH > UV > IR (infrared) > radiation*;
2. Dairy sector: *HPP > MWH > PEF > UV > OMH (omic heating)*;
3. Fresh produce: *UV >* O_3 *(ozone) > Radiation > HPP*;
4. Drink and beverages sector: *UV > HPP > PEF > OMH >* CO_2 *> MWH >* O_3 *> IR > power ultrasound*;
5. Fish and seafood sector: *HPP > MWH >* O_3 *> UV*.

It was not surprising that HPP and UV light scored the highest for all of these sectors. Comparing these two technologies, UV scored higher for fresh produce and drinks and beverages, while HPP scored higher for most of the food sectors, except for fresh produce. MWH scored better for all food sectors except for fresh produce. Ozone scored the highest for fresh produce, fish and sea products, and drinks and beverages. The existing regulatory approvals of radiation, HPP, ozone, MWH, and UV technology may be an important factor in their faster commercialization.

2.3 Production benefits

In addition to improving the quality and safety attributes of foods, novel processing technologies have the potential to reduce the three effects of thermal abuse, to lower the carbon footprint, and substantially reduce the water volumes and energy used in heat transfer processes. The international food community considers novel treatments as modern options to solve global challenges in addition to safety and shelf-life extension, especially in countries with a warm climate and often an underdeveloped infrastructure and cold chain. The level of processing technology development and regulatory status achieved in North America and Europe may accelerate their applications in developing countries.

3. Validation and verification as requirements of food safety regulations

3.1 New definition of pasteurization

The emergence of novel processing techniques using novel thermal and nonthermal physical and chemical high-technology methods and their approvals by regulatory agencies internationally has caused a need to review the definitions of traditional pasteurization and sterilization processes and to introduce new approaches for the establishment and design of the preservation process. In 2004, the USDA National Advisory Committee on Microbiological Criteria for Foods (NACMCF) redefined the term pasteurization (NACMCF, 2006). The rationale was that new physical methods of treatment are emerging as a result of the current consumer demands for foods that are more fresh, natural or minimally processed, and additive-free. As mentioned above, the use of thermal treatment can lead to overcooking, an undesirable cooked flavor, and nutritional deterioration. To satisfy these demands, traditional preservation techniques have been modified and novel processing techniques are used. Pasteurization was redefined as "any process, treatment, or combination thereof, that is applied to food to reduce the most resistant microorganism (s) of public health significance to a level that is not likely to present a public health risk under normal conditions of distribution and storage" (NACMCF, 2006). Cooking, steam and hot water treatments, microwave processing, ohmic/inductive heating, HHP, UV radiation, irradiation, PEFs, pulsed light, infrared processing, nonthermal plasma, oscillating magnetic fields, ultrasound, and filtration have been reviewed. Nevertheless, the efficacy of novel processing techniques in terms of inactivation of spores and spoilage endogenous enzyme inactivation is limited.

NACMCF also recommended *validating* the efficacy of the pasteurization process through the use of process authorities, challenge studies, predictive modeling, and/or safe harbors. All pasteurization processes that use novel technologies must be verified to ensure critical processing limits are achieved. While principles of establishment and validation of novel processes are under development it shall be considered that the same principles of microbial inactivation apply to other technologies based on thermal inactivation, such as microwave processing, ohmic heating, and surface pasteurization with steam. However, the mechanisms of microbial inactivation may differ for technologies that are not based on thermal inactivation, such as high pressure, UV light, and irradiation.

Also, it was recommended that a FSO and/or a performance criterion/objective (PO) has been proposed for use in establishing an appropriate level of public health protection for food pathogen combinations and could serve as the basis for equivalency when a proposed process is evaluated as an alternative to traditional pasteurization. A performance criterion is expressed by the Eq. (4):

$$H_0 - \sum R + \sum I \leq \text{FSO (or PO)} \tag{4}$$

where FSO is the food safety objective, PO is the performance objective, H_0 is the initial level of the hazard, Σ R is the total (cumulative) reduction of the microbial hazard on a $\log_{10}$ scale, and Σ I is the total (cumulative) increase of the hazard on a $\log_{10}$ scale.

In the case that the performance objective is measured immediately after treatment, the FSO shall be equivalent to the SLR (Eq. 1).

NACMCF recommended that regulatory agencies establish FSO and a performance standard for food pathogens that can be used as the basis for judging equivalency when a proposed process is evaluated as an alternative to traditional pasteurization.

NACMCF recommended considering numerous factors when establishing the efficacy and equivalency of different pasteurization processes. This includes identification of:

1. The most resistant pathogens(s) of concern in the food;
2. The efficacy of the specific technology to reduce the pathogens of concern;
3. The food matrix characteristics;
4. The intended use of food.

The modern market development trend shows that the pasteurized foods sector with extended shelf-life has become more prominent. This will continue because of consumer demand for products of high quality that can only be produced by applying minimal thermal processes. Over recent decades, a few novel intervention technologies have been successfully developed, validated, and accepted by regulatory agencies and applied as control measures to enhance the safety of minimally processed foods. The novel technologies undergoing research today comprise two general categories: (1) technologies suited for pasteurizing high-acid liquid products such as HHP, PEFs, ultrasound, UV light, and chemical processes, including gases; (2) technologies more appropriate for processing shelf-stable foods, for example, high pressure combined with temperature, microwave and radiofrequency heating, ohmic heating, and irradiation. For successful developments and commercialization of any novel processing technology greater understanding and evolving knowledge of the technology validation are critical in establishing the feasibility of a given technology for the purposes of preservation.

The number of publications on the validation of food processes is starting to increase. However, there are a few known areas where validation is an essential part of the product development and activities may have different content:

1. New applications for product formulations, manufacturing, and technologies to identify and eliminate food safety threats. These food processes must be validated and verified;
2. Sanitation including new chemicals, applications, and best practices for cleaning facilities. These products and practices must be validated and verified;
3. Technologies to detect pathogens, and indicator and spoilage organisms. These technologies must be validated and verified;
4. New laboratory methods and automation to turn around results faster, more accurately, and in a competitive low-margin market. These methods must be validated and verified.

The need for a clear understanding on the validation topic is paramount. The objective and focus of this chapter have been to review the definitions, principles, and practical approaches of *food process validation* and summarize the experience of thermal processing and a few commercially available novel processes that have been using HHP technology, and more recently UV light and other novel technologies for safety and preservation operations. The aspects explored include process applications and equipment; critical process, products, and control parameters for each technology; selection and recommendation of surrogates; and consideration for microbial validation studies and procedures used to measure and control the delivery of the required novel process.

References

Jermann, C., Koutchma, T., Margas, E., Leadley, C., Ros-Polski, V., 2015. Mapping trends in novel and emerging food processing technologies around the world. Innov Food Sci Emerg Technol 31, 14–27. Available from: https://doi.org/10.1016/j.ifset.2015.06.007.

NACMCF, 2006. Requisite scientific parameters for establishing alternative methods of pasteurization. J. Food Prot. 69 (5), 1190–1216.

US Department of Health, 1994. Guidelines for the Safe Production of Heat Preserved Foods. HMSO, London.

Further reading

Codex Alimentarius Commission. 2008. Guideline for the Validation of Food Safety Control Measures (CAC/GL 69–2008).
US Federal Register, 2013. Current good manufacturing practice and hazard analysis and risk-based preventive controls for human food. US Federal Register 78 (11), 3646.

CHAPTER 1

General principles and approaches of food process validation

Tatiana Koutchma[1] and Larry Keener[2]
[1]Novel Food Processing Technologies, Agriculture and Agri-Food Canada, Guelph, ON, Canada
[2]International Product Safety Consultants, Seattle, WA, United States

1.1 General principles and approaches of food process validation

1.1.1 Validation concept and terminology

Process validation is one of the core activities involved with establishing a process and demonstrating process and product safety and predetermined product quality and safety attributes. The adequate validation of any processing technology depends on more than just the "lethal or kill step" applied in the process flow. The entire food manufacturing system from receiving of ingredients and packaging material through to finished product storage and handling will ultimately define the Hazard Analysis and Critical Control Point (HACCP) or food safety plan for a product and will be reviewed by customers and regulators alike to assess the knowledge and care the food producer demonstrates when producing a product for the market.

Therefore there are multiple terminologies for "validation," "verification," and "monitoring" defined by regulatory and international organizations. It is important to keep all these tasks separate and to understand their role. While validation can be done using different approaches and many useful definitions are given, the endorsement is given to the use of Codex (CAC/GL 69-2008, Guidelines for the Validation of Food Safety Control measures).

US Food and Drug Administration (FDA) has required validation of pharmaceutical processes for over 30 years. Validation of pharmaceutical manufacturing processes has evolved over the years. The most recent documented pharmaceutical validation guidance was published in 2011.

Validation of Food Preservation Processes based on Novel Technologies
DOI: https://doi.org/10.1016/B978-0-12-815888-3.00006-3

According to the United States FDA (1987), process validation involves establishing documented evidence, which provides a high degree of assurance, that a specific process will consistently produce a product meeting its predetermined specifications and quality attributes/food safety.

There are three methods or approaches used for conducting process validation studies: prospective validation, concurrent validation, and retrospective validation.

Prospective validation is best suited for novel preservation technologies. This method is implemented during the design phase of a new processing method and aims to confirm the stability of the process for use in food production before any item of food manufactured using this method is safe before it is released for consumer use.

Concurrent validation is usually employed when a previously validated process or product has been changed or modified. The scope of this validation is to ascertain the impact of the change on the overall performance and stability of the process relative to established control limits. A new heater or pump or valve added to a process might be justification for a concurrent validation study. Likewise, changes in a product's formulation or a change in its packaging may also warrant a concurrent validation study. The advantage of this type of study is that the products manufactured under the validation protocols may be evaluated and approved for use in commerce.

Retrospective validation occurs when a previously validated process or procedure has failed. The study seeks then to retrospectively understand and identify the specific factors or other conditions that have contributed to the failure. Root cause analysis is an excellent example of retrospective validation.

The primary objective of validation is to demonstrate the product and specific process consistency over time across scales in research and development (R&D) and manufacture with various raw materials and operating ranges. *Process consistency* shows that the process, when operated according to manufacturing procedures, yields a product that consistently meets specifications. This type of validation usually is not required during early development. *Validation of operating ranges* illustrates that the process, when operated within established ranges for critical operating parameters, yields a product that meets released specifications. Documented evidence includes validation protocol of performance of the food processing system. The validation protocol is composed of the scope for the validation study and the detailed description of the procedures, methodologies, acceptance criteria, and responsibilities.

Acceptance criteria are governed by the predetermined specifications of safety and quality attributes, and they should be closely related to the risk of the process steps. For instance, as a result of disease outbreaks across North America associated with the consumption of unpasteurized juices and cider, the US FDA published a juice HACCP regulation designed to improve the safety of juice products [21 CFR Part 179 (United States FDA, 2000); HACCP (United States FDA, 2001)]. Under this rule, juice processors are required to produce juice under the HACCP system and achieve a 5-log reduction for the most resistant microorganism of public health significance that is likely to occur in the juice.

In FDA human foods regulations, the definition of validation includes control measure and food safety plan as a whole and determined as

> *obtaining and evaluating scientific and technical evidence that a control measure, combination of control measures, or the food safety plan as a whole, when properly implemented, is capable of effectively controlling the identified hazards.*
>
> ***United States FDA Federal Register (2015b)***

The Food Safety Modernization Act (FSMA) of 2015 reemphasized the importance of the validation and verification activities to achieve the safety of food systems. According to FSMA, validation is applied to the food safety plan, including some prerequisites, the hazard analysis, preventive controls, and corrective actions.

FSMA defines *validation* as an "element of verification focused on collecting and evaluating scientific and technical information to determine whether the Food Safety Plan, when properly implemented, will effectively control the identified hazards. FSMA clarifies "that proof can come from a variety of sources (e.g., scientific literature, in-house studies, mathematical modeling, and regulatory resources). An 'initial validation', also referred to as prospective validation, takes place as the food safety system is being developed and during its initial implementation. The goal is to ensure that the food safety system is valid (i.e., actually works) for controlling food safety hazards associated with the operations, ingredients, process, and product." Prospective validation seeks to prevent the introduction of foods into the marketplace that may contain food safety defects at levels exceeding public health thresholds.

Performance validation is an essential part of the evaluation of new process suitability as an alternative preservation method. All pasteurization processes need to be validated through the use of process authorities, challenge studies, and predictive modeling.

National Advisory Committee of Microbiological Safety of Foods (NACMCF) along with a new definition of pasteurization process, defined *validation* as the collection and evaluation of scientific and technical information to determine if the treatment, when properly applied, will effectively control the microbial hazard(s) (NACMCF, 2006). For validation, an expert in the technology area needs to determine the critical factors for each application. Product and process variability, batch-to-batch variation, influence of product characteristics, size of product pieces, etc., need to be considered. All pasteurization processes must be verified to ensure that critical processing limits are achieved. As novel technology is applied commercially, the research is needed to develop label statements about pasteurization that can be understood by consumers.

Validation can be done using different approaches and many useful documents have been published. The support and endorsement are given to the use of Codex (CAC/GL 69-2008, Guidelines for the Validation of Food Safety Control measures) because this is an international standard that is widely recognized throughout the world. Codex also provided definitions of process validation as "obtaining evidence that a control measure or combination of control measures, if properly implemented, is capable of controlling the hazard to a specified outcome." *Control measure* is any action and activity that can be used to prevent or eliminate a food safety hazard or reduce it to an acceptable level. In other words, a control measure can be any unit operation (thermal or nonthermal) that results in control or reduction of the target organism. This includes the actual transport phenomenon related to treating the food product (Table 1.2) and the physical or chemical characteristics of the food, or treatment, that affects or augment the dynamics of the process. Within each Control Measure there will be Control Points and/or Critical Control Points (CCPs) that will need to be regulated and monitored for the Control Measure to be effective. Validation serves to confirm the critical limits operational at the CCP(s). Critical limits define the accepted level of risk for a specified process step or procedure. The effectiveness of a process can be measured and monitored in terms of its performance relative to the established critical limits. Excessive variation in the critical limits, exceeding the upper control level, is an indication of an unstable and unreliable process. In terms of food safety, an unreliable process is a potential threat to public health. The concept of food safety objectives (FSOs) and/or performance criteria (PC) has emerged in food safety science as enablers of the validation process. Recall that validation by definition is targeted to a predetermined specified outcome. The FSO provides guidance in establishing the accepted level of risk for an item of food manufactured by a specific process. Setting the specified outcomes (public health) has long been the bane for those writing validation protocols.

Codex document also identifies a few approaches for process validation including reference to scientific or technical literature, previous validation studies, or historical knowledge of the performance of the control measure; scientifically valid experimental data that demonstrate the adequacy of the control measure; collection of data during operating conditions in the whole food operation, mathematical modeling, and surveys. The selection of the approach or combination of the approaches will depend on the nature of the hazard, the nature of the raw ingredients and product, the type of control measures, the history of its use and available scientific information and food safety control system selected to control the hazard, and the intended stringency of control of the hazard.

1.1.2 Verification versus validation versus monitoring

Verification is often can be confused with validation. Two aspects should be considered when applied to verification: verification of control measure and verification as a critical component of a food safety plan.

Starting with the basics of verification of control measure, verification can be defined as a procedure by which it is confirmed that a control measure or process is operating as designed. Verification is an ongoing procedure that monitors the validated process while the product is produced.

United States FDA (2015a) defined verification as "the application of methods, procedures, tests and other evaluations, in addition to monitoring, to determine whether a control measure or combination of control measures is or has been operating as intended and to establish the validity of the food safety plan."

Verification activities have been defined by FSMA as "those activities, other than monitoring, that establish the validity of the food safety plan and ensure that the system is operating according to the plan."

A similar description of verification is proposed by Codex document, where verification is defined as "the application of methods, procedures, tests and other evaluations, in addition to monitoring, to determine whether a control measure is or has been operating as intended." Because verification cannot be interchangeable with validation, Codex also added essential clarification in order to avoid confusion among the concepts of validation, verification, and monitoring, Specifically, Codex states that "monitoring and verification both take place after the validated control measures have been implemented. Monitoring and verification are the tools used to check whether the control measures are being adhered to and to demonstrate that they are operating as intended." Monitoring of control measures is the ongoing collection of information at the step the control measure is applied. The information establishes that the measure is functioning as intended, that is, within established limits. Monitoring activities are typically focused on "real-time" measurements and on the performance of a specific control measure. The following examples of monitoring include measuring pH drop during fermentation, weight loss [or water activity (A_w)] during maturation, and temperature measurement during the heating process.

Verification occurs during or after the operation of a control measure through a variety of activities, including observation of monitoring activities and review of records to confirm that implementation of control measures is according to design. Examples of verification activities include periodic process control testing for pathogenic *Escherichia coli* to verify that incoming levels in the raw materials are within specification and that fermentation and maturation achieve the intended outcome in the semi-finished or finished product, and examination of monitoring records to check for continuous control over time.

1.1.3 Validation at different phases of process development

A critical component of process validation is the selection and qualification of scale-up models. Three important aspects of the scale-up models that include design and development, performance, and product quality must be considered during qualification before process validation experiments are begun.

Design refers to the use of scientifically valid principles in specifying conditions for the scale-up process. For example, all the materials that are used for the equipment must meet the US FDA Food Equipment Contact Surface Requirements and principles of hygienic design. *Performance objectives* of the various scale-up models must be comparable, and the differences among the models have to be quantified. For instance, in ultraviolet (UV) light-based systems, the characteristics of the UV light source, geometry of the chamber, flow pattern, and mixing characteristics through the use of dimensionless numbers need to be considered as critical in the scale-up process. The models must have sufficient sensitivity to detect changes when input variables are manipulated. Finally, the *quality* aspect calls for identical quality among the various scale-up models where any differences among the models must be quantified.

1.1.4 Scale-up process

The four phases of validation activity are common during the scale-up process of any new technology. R&D of a new food process typically includes activities during four phases of bench top, pilot model, prototype skid to the commercial processing system (Table 1.1). The probability of process success has to be evaluated step by step during the whole validation process.

Bench-top development in the laboratory includes the development of the principal process scheme, the definition and validation of process operating parameters, the inactivation kinetic studies with well-controlled and measured parameters, the establishment of the processing parameters and product attributes for required log reduction on the pertinent pathogen of concern, and the determination of shelf life and packaging requirements. Also, suitable surrogates that imitate inactivation and growth behavior of pathogenic organisms in food products have to be selected during bench-top development. At the end of this phase, a process has to be defined for the production of a high-quality product by the bench-top model.

As an example, tests using collimated beam apparatus, UV chambers, continuous bench-top UV systems with well-controlled and measured UV parameters can be used in UV process development for liquid and solid products and should provide repetitive consistent results. Biodosimetry studies using collimated beam apparatus that are conducted to determine a

Table 1.1 Validation activities on different phases of process scale-up.

Phase	Process validation				Equipment	Analytical	Cleaning
	Microbial	Physical	Quality sensory shelf life	Packaging			
Bench top	X	X	x				
Pilot	X	X	X	X	X	X	
Prototype	X	X	X	X	X	X	X
Commercial	X	X	X	X	X	X	X

reduction equivalent UV dose for a pathogen of concern is the first step in process validation (Kuo et al., 2003).

The *pilot model* is an important step during the scale-up of a production process. It requires formal qualification of the equipment and the appropriate calibration of instruments. Process validation usually concentrates on the scalability and reproducibility of the process and product quality. If necessary, process modifications should be made to optimize the production process for full-scale manufacturing. Products derived from pilot and prototype production should possess high-quality attributes and be safe (hazard free).

On-site validation is preferred in the food industry. Equipment that will be used for *commercial-scale* production has to be qualified according to Good Manufacturing Practices (GMPs) and current industrial standards, and an appropriate calibration of the instrument is required. Process validation should demonstrate that the product produced by the commercial unit would be comparable to that produced by the pilot/prototype models and that the commercial unit can be operated within the predetermined process parameters and specifications and food product variation.

The first criterion is to measure and document the performance of the commercial UV unit at established processing parameters: X (flow rate), Y (processing time), and Z (UV dose). The production efficiency must be calculated based on the time for production, clean-in-place (CIP) and sanitation-in-place requirements, and inspection time. The second criterion is to meet the required inactivation at 95% confidence levels. Variances associated with measurements made during validation are quantified at the 90% confidence level. The third criterion is to demonstrate a consistent performance over 4 consecutive weeks of 24 hours/day to 5 days/week operation.

In addition, a well-designed cleaning program and analytical methods established for in-process control have to be fully validated. The food processors should follow 3-A sanitary standards and 3-A accepted practices or Canadian dairy standards for their equipment. Equipment manufacturers, fabricators, users, and sanitarians universally accept 3-A criteria. Sanitary Standards of 3-A (3-A SSI) formulates standards and practices for the sanitary design, fabrication, installation, and cleanability of dairy and food equipment or systems used to handle, process, and package consumable products where a high degree of sanitation is required. Individual standards are now available in downloadable electronic format at http://www.3-a.org/main.htm.

1.1.5 Key components of validation procedures

Process validation can be accomplished by microbiological and/or physical methods and includes evaluation of product quality. Therefore the key components for the validation procedures of any new process include validation of process uniformity, effect of critical process parameters, microbiological and chemical safety, product quality, equipment performance and calibration, facility, cleaning, and analytical elements as shown in Fig. 1.1. Each element has its own objective(s) and procedures.

1.1.6 Physical validation

Physical validation means that the critical process parameters must be measured within the processing range. The question to be answered during physical validation is if the process can deliver a uniform treatment between food units with respect to specific process parameters, such as temperature, pressure, concentration, time, X-ray, or UV doses absorbed (i.e., the processing parameters that affect the treatment dose).

In thermal process validation, one of the core activities is the measurement of product temperatures that represents the "worst-case conditions" likely to be experienced during normal production. For in-container heat processing, this activity typically involves two stages of temperature measurements: temperature distribution or TD tests to identify the location of the zone of slowest heating in the retort, and heat penetration or HP tests to measure the temperature response at the product cold point. For a continuous flow process, a specialized form of the HP test is required, with

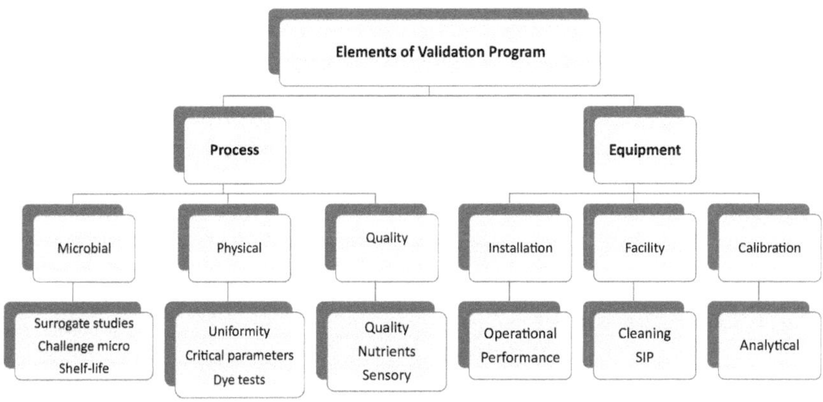

Figure 1.1 Chart of validation procedures of a new process.

various assumptions made regarding the positions of temperature measurement that make the TD test redundant.

In the validation of UV processing of food liquid or beverage—the location of the least UV-treated zone in the product must be determined in addition to the product flow rate and other flow parameters such as residence time distribution. Also, the characteristic properties of the product that can impact the efficacy of the process have to be measured and the range of their variation has to be estimated. This includes parameters such as pH, A_w, viscosity, and absorption coefficient. It is also beneficial to measure the composition of the product and evaluate the potential impact of the essential components on process delivery like the content of ascorbic acid that has high UV-absorbing properties.

1.1.6.1 Process uniformity evaluation

Any processing system will contain regions in which the value of physical lethal agent (temperature, UV intensity, pressure) is lower than the mean average parameter measured by master instruments, such as intensity sensor, thermometer, or pressure sensor. The location of these cold spots should be determined by performing distribution tests throughout the system. The concepts for distribution testing are simple; however, the practicalities of making the relevant measurements can be filled with difficulty. For solids foods that heat mainly by conduction, the slowest heating point will be at the container's geometric center. However, for foods that permit movement and include convection heating, this point is between the geometric center and approximately one-tenth up from the base (in a static process).

Calibration of sensors against a traceable instrument is essential each time they are used in distribution trials. Another approach for conducting this kind of special distribution measurements is using time–temperature integrators, pressure sensors, special sensitive dyes, or actinometers. Additionally, the advances in microchip technology can provide more resources in measurements and analyzing the results, and increase the accuracy.

1.1.7 Microbiological validation

1.1.7.1 Challenge studies

1.1.7.1.1 Pertinent pathogen(s) selection

In any process validation study, the target pathogen of concern must be identified. According to the juice HACCP regulation (21 CFR 120), this microorganism is called the "pertinent pathogen." There is guidance in the Juice HACCP Alliance's standardized training curriculum and US FDA's Hazards and Controls Guide to aid this process.

Knowledge of the food formulation and history of the food, for example, association with known illness outbreaks and/or evidence of potential growth is essential when selecting the appropriate challenge pathogens. The ideal organisms for challenge testing are those that have been previously isolated from similar formulations. Additionally, pathogens from known foodborne outbreaks should be included to ensure the formulation is robust enough to inhibit those organisms as well. Multiple specific strains of the target pathogens should be included in the challenge study. It is typical to challenge a food formulation with a "cocktail" or mixture of multiple strains in order to account for potential strain variation and differences in sensitivity. In challenge studies such as this, a "cocktail" of at least three to five strains of the pertinent pathogen would be selected, based on previous association with the product or process in question (e.g., outbreak strains associated with a particular product). Use of a cocktail in this way gives assurance that the strain most resistant to the challenge would be the one that survives to be evaluated, and hence would present the most conservative picture, at least experimentally, of the efficacy of the process. Clearly, for any particular process there may be more resistant strains yet to be discovered in the environment.

It is not unusual to have a cocktail of five or more strains of each target pathogen in a challenge study. For example, *E. coli* O157:H7 has been recognized as a cause of serious illness and mortality in outbreaks when unpasteurized apple cider or apple juice was consumed. Shiga toxin-producing *E. coli* (STEC) serogroups: O26:H11, O103:H2, O145:NM, and O111:NM also were recently recognized as emerging human pathogens and were associated with several juice outbreaks (Orlowska et al., 2014). Non-O157 STEC serogroups cause indistinguishable O157:H7-induced disease and therefore research was mainly focused on *E. coli* O157:H7.

There may be even more resistant spoilage (i.e., not safety-related) organisms that may have to be considered as a final preservation process is established. However, the use of target pathogen strains in studies is a conventional approach, which, in a properly designed and executed study, will meet the need to demonstrate the efficacy of the process through inactivation of a target level of the pertinent pathogen according to the requirements of the juice HACCP regulation.

Once the pertinent pathogen(s) is (are) identified, a thorough search of literature, bench-top studies, or discussions with experts will identify gaps in knowledge with respect to the level of treatment to yield a required 5-log inactivation of the microorganism using any particular processing technology. At this stage, various laboratory studies may be designed and performed to generate new data to close the gaps in knowledge.

Good experimental design will consider the number of replicate studies needed to provide data that can be analyzed for statistical validity, the number of variables to consider in such a design (i.e., the number of multiple treatment parameters), and the choice of adequate controls. Based on the most appropriate design, the ability to effectively provide a statistical evaluation of data, for example, to determine statistical significance of results with respect to the first-order or linear inactivation model is determined. This relates to integrated lethal processes applied to apple juice, and the cumulative reductions that are part of the stepwise interventions allowed in the juice HACCP rule applied to whole citrus fruits.

1.1.7.2 Microbiological methods

One important consideration regarding the choice and application of challenge strains is the need to "adapt" them to their most resistant level. Obviously, in the wild or in the manufacturing environment, the would-be microorganisms of concern most likely would not exist in ideal conditions. Very often they survive, but maybe "stressed" and very closed to death, under adverse conditions such as low environmental pH or near-starvation. So, it is important to "stress-adapt" the strains used in inoculation studies, under conditions, which simulate as much as possible the conditions imposed in the future challenge. Prior culture, for example, in nutrient-limiting (starvation) conditions or other nonoptimal growth conditions such as low pH, high or low temperatures, reduced A_w, or others, either alone or in combination with other adverse pretreatments, is often done.

For example, acid adaptation of *E. coli* O157:H7 cells or *Salmonellae* cells prior to inoculation can greatly influence their ability to survive when inoculated into an acidic food. When *E. coli* strains were kept in apple juice at low pH 3.5 tested strains up to 24 hours at room temperature (22°C), the bacterial population of *E. coli* O157:H7 remained unaffected by the acidic environment. The ability of *E. coli* O157:H7 to withstand well the high-acid conditions was the cause of several outbreaks related with the consumption of unpasteurized apple juice/cider.

1.1.7.2.1 Inoculum Levels

Generally, the extremes of contamination levels are explored (i.e., high vs low inoculation levels). Although generally artificial with respect to the "normal" levels seen on fruit or in juice processing operations, use of the former (e.g., $>10^6$/mL or g) can be applied under appropriate treatment conditions to give a measurable level of survivors which facilitate comparisons of the effects of different process variables.

Since high bacterial concentration levels can influence absorbance of the sample solution, the appropriate maximum initial inoculation level for the inactivation studies needs to be determined. The absorbance coefficient needs to be measured and the rates of inactivation at various bacterial inoculation levels are compared. It was found that initial inoculation levels of up to 10^7 cfu/mL had no effect on the absorbance of the sample solution. However, the absorption coefficient of the sample tended to increase when the initial count was higher than 10^8.

The lower inoculum levels, while reflective of more "typical" anticipated levels found in the process, will provide lower levels of survivors under equivalent processing conditions. In these instances, specialized techniques (e.g., evaluation of large volumes of product combined with most probable number or other sensitive techniques) must be used to enumerate survivors. A comparison of studies performed with high and low levels of inoculum will determine if uncontrolled process effects produce key differences in predicted levels of inactivation. For routine enumeration of survivors, it is traditional to recover survivors on both selective and nonselective media. The difference in numbers counted between the two media will give a measure of the number of sublethally injured microorganisms surviving the process. The total count from the nonselective medium will provide the basis to evaluate process PC.

The exact inoculation method is critical. Of course, to the best of available knowledge, this should effectively resemble the most likely route of contamination or ingress of the pathogen into the product. This is most important with respect to surface inoculation of fresh fruits or vegetables before pressing or extracting the juice. This particular consideration is presently an area of considerable research activity, although some information regarding optimization of the inoculation process is currently available.

1.1.7.2.2 Model systems

Usually, inactivation models can be developed in media designed to simulate the natural product as closely as possible, since the use of such media is very often much easier to work with than the natural product. The models are then validated in the actual food, usually using a smaller number of experimental variables, which is another benefit that the simulated modeling route often permits. With juices, which are generally easy and inexpensive products to work with, the need to develop models in simulated product media may not be a paramount consideration; that is, inoculated studies very often are performed using the product in question directly. Finally, it is usually customary to monitor the viability and potential for the growth of surviving microorganisms throughout the anticipated shelf life of the product. It is conceivable, but highly unlikely given the low pH of most juice products that sublethally injured microorganisms may repair to full viability and that fully viable survivors may actually increase in numbers with time.

However, in order to reduce the effects of variation of fresh juice properties, juice simulants can be used in the initial stages of validation of UV light processes. Koutchma et al. (2005) used model caramel solutions to simulate absorption properties of exotic tropical juices and to study the effect of absorbance on UV inactivation since the cost of shipping of the required volumes of juices from Hawaii was very high. The base for the model system was a 0.025 M sodium malate buffer at pH 3.5 and sucrose was added to 10 °Brix. Caramel (Pepsi Cola Co, New York) was added in the range of concentration from 0.2% to 2.0% to approximate tropical juices.

The absorbance of a 0.2% model caramel solution was comparable to the absorbance of a clarified lillikoi juice. The absorbance of a 0.5% model caramel solution matched with that of unfiltered apple juice. The absorbance of model caramel solutions of between 1% and 1.1% was similar in absorbance of commercial orange, guava, and carrot juices, and 2.0% model caramel solution was used to mimic the absorbance of pineapple juice.

1.1.7.3 Microbial validation in process scale-up and use of surrogate organisms

Once processing parameters to inactivate the target level of pertinent pathogen have been established in laboratory studies, the process is usually scaled up through pilot plant development and then through prototype equipment and process development concluded by commercial rollout and commissioning of the system.

Obviously, unless working in a controlled environment (e.g., a Biosafety level, or SL, two or three pilot-scale containment facilities), pathogens should not be used in an open processing environment where there is a potential for contamination of equipment and facilities. In these situations, nonpathogenic "surrogate" organisms or challenge microorganisms are selected for use in further studies in lieu of the pertinent pathogen. The use of surrogate microorganisms is a valid approach when an appropriate biosafety laboratory is not available and because it is not acceptable to introduce pathogens into a food processing facility.

Only a few studies on thermal sterilization processes use *Clostridium botulinum* spores because of the hazards associated with their handling, but also due to the very low number of surviving spores that would result from a commercial process. Hence, a nonpathogenic organism with a high $D_{121.1}$ value is used, such as spores of *Clostridium sporogenes* or *Geobacillus stearothermophilus*. Typical levels of the inoculum are between 10^3 and 10^5 spores per container. Montville et al. (2005, 2006) identified avirulent *Bacillus* spores as an appropriate surrogate for the virulent *Bacillus anthracis* in thermally processed milk and orange juice. Eblen et al. (2005) identified *E. coli* ATCC 25922 as a candidate surrogate for *E. coli* O157: H7, and *Salmonella poona* for treatment of apples with hydrogen peroxide.

Obviously, such challenge strains must have resistance traits that have been predetermined in controlled studies to match as closely as possible with those of the pertinent pathogen. Moreover, many of them will have a marker (e.g., natural antibiotic resistance) which will facilitate their identification as process survivors separate from any number of similar strains found as indigenous flora in the natural product. Other characteristics desirable for chosen challenge strains have been described by IFT (2000). Very importantly, the chosen surrogate (challenge) strain, if it survives the process, should not itself become a product spoilage organism and will not persist as a hard-to-remove nuisance organism in the processing facility or the process waste stream.

Suggested levels of microbial validation at four phases of process scale-up are summarized in Table 1.2.

If a surrogate strain is to be used in a microbiological validation study, preliminary work should be conducted to well characterize the strains in the laboratory before using for the prototype or commercial validation study.

1.1.7.4 Criteria for selecting pathogen surrogates

According to the IFT (2000), the surrogate microorganism is defined as "non-pathogenic species and strain responding to a particular treatment in

Table 1.2 Levels of microbial and quality validation at different phases of scale-up.

	Pathogen	Surrogate	Spoilage	Quality
Concept/laboratory	X	X		X
Pilot	X	X		X
Prototype		X	X	X
Commercial		X	X	

a manner equivalent to a pathogenic species and strain." Following criteria and steps for the selection of the surrogate organism should be considered to validate process:

1. Surrogates are nonpathogenic to testing personnel if handled properly. According to Hu and Gurtler (2017), surrogates are generally selected from a population of well-known organisms that also have well-defined characteristics and a long history of avirulence and can easily be differentiated from other microflora.
2. Phylogenetically close to the pathogen of concern capable of coexisting with the pathogen in the food matrix.
3. Surrogates should be easy to prepare. The least affected by the nature of food product (pH). The survival of pathogen and nonpathogenic strains has to be tested in the food matrix (pH, A_w).
4. Screening studies using pathogenic and nonpathogenic strains selected in step 3 are conducted in parallel in order to determine nonpathogenic strains characterized by equivalent or higher resistance to the lethal agent than pathogens in a given food product.
5. Inactivation studies are conducted in the product inoculated either with single strain or cocktail, that is, mixture of pathogen and nonpathogenic strains, which were selected in step 4. The aim of the step 5 is to compare inactivation curves of the surrogate candidates with pathogen. Detailed description of the processing conditions should be provided.
6. Fit the experimental data with the correct models to obtain kinetic parameters that can be used in commercial operations.
7. Surrogates should have stable and consistent growth characteristics including the growth of the bacterial residuals in the posttreated food product.

Criteria such as those discussed above should be determined and confirmed through available reported data and preliminary laboratory work to assure that the surrogate strain is suitable for the intended purpose and process. The choice of a surrogate in a process validation study must be justified with supporting documentation with regard to the target pathogen, food, and intervention treatment being evaluated. All parameters should be explained in detail in the validation study report.

The selection of surrogates has been reported for common food pathogens of concern such as *Listeria monocytogenes*, *Salmonella*, *E. coli O157:H7*, and *C. botulinum*. Hu and Gurtler (2017) provided a list of validated surrogates strains reported in the literature and that can be recommended instead of the corresponding pathogenic strains.

Listeria innocua strains are considered as attractive options as surrogates for *L. monocytogenes* because of their high genetic similarity and innocuous nature. *L. innocua* ATCC 33090 is commonly used as a surrogate for *L. monocytogenes* in studies involving ozone, smoking, vacuum packaging, heat pressurization, and irradiation in various types of food products. *L. innocua* M1 (ATCC 33091) was developed by Fairchild and Foegeding (1993) as an antibiotic-resistant surrogate for *L. monocytogenes* to study the effects of thermal processing in milk. However, other recommendation was that carefully chosen *L. innocua* strain might be an effective surrogate for *L. monocytogenes* in controlled laboratory studies, but might not be the most appropriate option as a surrogate for *L. monocytogenes* in a food production facility or in a pilot plant where products may eventually be produced for human consumption.

E. coli ATCC 25922 and *E. coli* K-12 strains have been popular choices as surrogate strains in place of *E. coli* O157:H7 in food challenge studies. In several studies, *E. coli* K-12 strains have been used as surrogates for *Salmonella enteritidis*. Gurtler et al. (2010) reported sensitivity screening of 23 pathogenic and nonpathogenic strains of bacteria, including *E. coli* ATCC 25922 and a wild-type *E. coli* K-12 (ATCC 23716), for treatment using pulsed electric field (PEF) in orange juice. The wild-type *E. coli* K-12 was the most sensitive of all nonpathogenic strains and should not be considered as a surrogate for either *Salmonella* or *E. coli* O157:H7 in PEF treatment of acidic juices and beverages because of its high sensitivity. Also, the authors validated nonpathogenic *E. coli* ATCC 35218 as the bacterium that characterized by most close inactivation behavior of the pathogenic *E. coli* O157:H7 and *Salmonella* strains and was only slightly more resistant to PEF treatment.

Salmonella typhimurium LT2, antibiotic-resistant ATCC 700720 strain, is commonly used as a surrogate for *S. typhimurium* in published studies and is naturally void of the rpoS operon, hence rendering the strain nonpathogenic or less pathogenic than the wild-type strain. It was recommended as a potential surrogate for the validation of UV processes and irradiation.

Enterococcus faecium NRRL B-2354 is a surrogate bacterium recommended and commonly used in place of *S. enteritidis* PT 30 for the validation of thermal processing technologies used for almonds and pet foods.

C. sporogenes PA 3679 is a long-time accepted surrogate for *C. botulinum* when used in inoculated pack studies to validate thermal processes for low-acid canned foods but cannot be used as a direct substitute to validate a product for inhibition of botulinum toxin production. The heat resistance of proteolytic *C. botulinum* is reported as significantly lower than that of *C. sporogenes* PA 3679. It was also validated for the use in pressure-assisted thermal sterilization (PATS) processing.

Bacillus cereus and *Bacillus subtilis* spores are used as surrogates for *B. anthracis* spores in food, water, and environmental intervention studies, especially when attenuated *B. anthracis* strains are not permitted in certain laboratories.

A few evaluated and reported recommendations of surrogate nonpathogenic organisms for thermal and nonthermal processing of some foods and beverages are summarized in Table 1.3. More detailed list of recommended surrogates can be found in the review by Hu and Gurtler (2017).

1.1.8 Quality validation

1.1.8.1 Quality, nutritional content, shelf-life and sensory studies, packaging

Besides ensuring the absence of pathogenic microorganisms, inactivating indigenous microorganisms and enzymes that are responsible for spoilage and quality deterioration is another critical goal in food processing. For example, the quality of juice is defined by the indigenous microbial populations, indigenous enzyme activities, physical attributes (pH, Brix, color, viscosity, and cloud), chemical attributes (sugar profile, acid profile, minerals, and titratable acidity), nutritional content (ascorbic acid, folic acid, and beta-carotene) and organoleptic properties (aroma profile, flavor profile, and texture).

Shelf-life studies are commonly performed to assess the consistency of the treated product quality throughout its targeted shelf life. During the shelf-life study, samples are periodically tested for indigenous microorganisms, indigenous enzymes, and the evaluation of other quality attributes.

Table 1.3 Surrogate organisms for the validation of some thermal and nonthermal processes.

Pathogen	Nonpathogenic surrogate	Product	Process	Reference(s)
Listeria monocytogenes	*Listeria innocua* ATCC 33090 *L. innocua* ATCC 49594	Brain heart broth CodOrange juice Range of foods Endive leaves	Heating at 60°C High-pressure processing Irradiation	Lorentzen et al. (2010) Char et al. (2010) Niemira (2010)
Escherichia coli O157:H75 strains	Lactic acid bacteria	Beef jerky butter	Heating	Borowski et al. (2009a)
Salmonella enterica 10 strains	Lactic acid bacteria	Beef jerky butter	Heating	Borowski et al. (2009b)
Salmonella serovars *Typhimurium* ATCC 14028 *Typhimurium* UK	*Salmonella typhimurium* v3985, v3552, v4096; *E. coli* ATCC 29994, 35270, 35218; and *Lactobacillus casei* ATCC 393	Orange juice	Pulsed electric fields	Gurtler et al. (2010)
Clostridium botulinum	*Clostridium sporogenes* PA 3679	Low-acid foods	High-pressure thermal processing	Ahn et al. (2007)

In addition, factors such as types of packaging or packaging materials (glass, plastic, paper, carton), storage temperature (abused and normal cold chain distribution refrigeration temperature), and transportation should also be included in the study. Standard plate counting techniques are generally used to estimate the total number of viable cells in the treated products during storage. The yeast and mold populations and lactic acid bacteria counts are also determined, as these are microorganisms that are principal in spoilage.

Pectins are a group of colloidal carbohydrates naturally present in fruit juices. They act as natural stabilizers and give the fruit juice a consistency that is generally referred to as "body" by the fruit juice industry. Pectinesterase is the indigenous enzyme that degrades the pectin colloid and decreases the cloud stability of the juice. The titration method using a pH stat system or the colorimetric method developed by Rouse and Atkins (1955) can be used to measure the activity of pectinesterase or Personal Protection Equipment (PPE) in citrus juices. Standard tests such as the official methods from Association of Official Analytical Chemists (AOAC) or the Department of Citrus, Florida, must be employed to analyze the physical, chemical, and nutritional attributes of the juice.

Sensory evaluation by expert and consumer panels is used to document changes in sensory profile and consumer acceptance. Analysis of the aroma profile by gas chromatography is employed to supplement the organoleptic information collected from sensory evaluation.

1.1.8.2 Chemical and physical safety

During process validation, potential chemical and physical hazards must also be addressed. For example, the production of toxic by-products and the use of unapproved additives and inadvertent introduction of potential allergens must be avoided by proper process design and evaluation. Likewise, introduction of physical hazards such as contamination of product by introduction of foreign objects during processing must be prevented. Examples include introduction of glass fragments from bottles as a result of cracking through inappropriate temperature cycling, or the inclusion of metal shards from lids as a result of exceeding tolerance in filling operations. Again, properly designed and executed process validation studies will effectively consider the likelihood of such occurrences and lead to measures for their prevention, elimination, or reduction to acceptable levels.

As there can be uncertainty about the production of potentially toxic substances such as free radicals by novel processes, it is essential to confirm the absence of these unwanted by-products as part of the validation program. Hence, prior to commercialization, open dialog among regulatory agencies, juice processors, and technology developers must be initiated to address the labeling and grading issues for these new categories. Last but not least, variations due to the properties of raw materials (e.g., beginning/end of season fruit, varieties of fruit, amount of pulp) should also be analyzed and established.

1.1.9 Equipment validation

The key elements of equipment validation include initial inspection, installation and operational qualification, performance qualification, and commissioning. Prior to the installation of the equipment, a description of the equipment, checklist of the software, and accurate supporting documentation must be provided. The manufacturer of food processing system must provide information on the proper installation, operation, and maintenance. For UV light processing system, this includes performance characteristics, such as nominal pressure, flow rates, head loss, pressure rating of the sleeve, UV intensity sensor and monitoring window used; assembly and installation instructions; electrical requirements; and operation and maintenance manuals. During *installation and operational qualification*, sensors need to be calibrated against standards (e.g., National Institute of Standards) and control loops have to be tuned.

The *performance* of the system must be validated over the operating range of processing conditions for each specific product. The following approaches can be used to estimate *performance qualification for* "the worst-case scenario approach" or "the process steps approach." Inoculated pack studies can be run for the selection of the appropriate microorganism. The worst-case approach is attractive since the validation can be accomplished in only a few experiments if they are successful. For efficient analysis, the sampling plan and statistics must be used along with optimal experimental design. The process steps approach requires the establishment of the processing steps prior to the validation of the system performance. The set point values of process variables should be determined from process development studies. In practice, the exact control of a variable at the set point is seldom achieved and the variable is typically maintained within the range of values or normal operating range (NOR). The magnitude of the NOR can be ascertained during a production trial and set at 2 or 3 standard deviations. To cover occasional excursions outside the NOR, it is desirable to establish a wider range known as the maximum operating range or MOR within which product quality attributes have been shown to be acceptable. Any performance outside the MOR is at the edge of failure.

The objective of process validation is not to determine the edge of failure but to demonstrate acceptable process operation. In juice validation studies, processing parameters must be varied to get (+)/(−) versus microbial inactivation. To ensure that all system components are functioning, they need to be challenged on process deviations as well. Another key element of the system performance evaluation is to demonstrate the uniformity of the process. Supporting documentation should describe all aspects of the system that impact performance parameters.

At the *commissioning phase*, an equipment validation system acceptance test should be run. First, normal CIP and sanitization steps on the system should be performed, then environmental air samples are collected and swabs in the critical areas should be taken. The product is passed through the system using the designed process. The samples are collected according to a statistical sampling plan at the beginning, middle, and end of the production for microbiological counting. It is highly desirable to repeat the test three times and to collect samples from each trial. In order to accept the system for production, the defective rate must be less than 1%.

1.1.10 Cleaning and sanitation validation

Cleaning strategies adopted in multiuse facilities play a major part in the prevention of cross-contamination of food products. Government-approved chemicals must be used for cleaning and sanitation. Cleaning different pieces of equipment has to be evaluated on a case-by-case basis and justified according to performance qualification studies. Cleaning routine must be defined, validated, and set. The three main parameters that are used for the evaluation of equipment cleanliness are analysis of final rinse water samples at the end of cleaning, analysis of surface swab samples, and visual inspection. The buildup of biofilms has to be considered in cleaning validation.

In order to pass an equipment sanitation test, the validation test must be performed three times. In case of fresh juices, juice is inoculated with mold spores and circulated through the system for 30 minutes. Then the system is drained, dried, and cleaned with normal CIP and sanitation cycles. After cleaning, the machine is disassembled and swabbed in preselected sites. Microbiological testing is then performed on the swabbed samples using standard procedures. Dye tests can be also used for determining cleaning effectiveness.

Automated CIP can be used and validated against established cleaning regimes. The juice processors can consider two options for CIP validation. The first method includes circulation of buttermilk in the system overnight, drying of the system for 4 hours, running normal CIP cycles at proper conditions, adding erythrosin dye during the final rinse, and at the end disassembling of the machine for the diagnosis of dye residue. The second option involves circulating riboflavin in the system, drying the system for 4 hours, running the normal CIP cycles, and disassembling the machine for the diagnosis of florescent residue. European Hygienic Equipment Design Group may have other methods for CIP validation to assist industry in complying with European hygienic machinery directives.

1.1.11 Facility requirements

Manufacturing facilities require key operational strategies to be in place. Safety and worker safety do not need to be part of food safety validation but could be considered as a part of the overall plant validation program. Some of the basic requirements for a manufacturing facility include well-designed facilities and equipment, appropriate personnel, product, raw materials, and waste flows, and training programs for personnel with regard to cleaning. A cleaning validation program specifically for facilities and equipment is required for all GMP-regulated food facilities.

Also, virtually every food facility, both in the United States and abroad, that is subject to FDA's Bioterrorism Facility Establishment registration must establish and implement an adequate HARPC food safety plan. HARPC stands for "Hazard Analysis and Risk-Based Preventive Controls." As required by FSMA, FDA promulgated the HARPC regulation (also referred to as the "Preventive Controls Rule" for the food industry (United States FDA Federal Register, 2015b). The owner, operator, or agent in charge of the facility must have their HARPC plan available and ready to present to FDA.

HARPC requires food facilities to design and implement verification steps to ensure that their HARPC plans including the hazard identification and analysis, preventive controls and control measures, monitoring and corrective action steps are operating correctly to prevent or minimize food safety and adulteration hazards. Verification steps should be sufficiently robust to ensure that:

- The selected preventive controls are adequate.
- Monitoring is occurring properly as defined in the plan.
- Appropriate corrective actions are taken.
- Potential food and food processing hazards are reduced.
- Periodic reviews are conducted at appropriate intervals so the HARPC plan remains working and takes into account new and emerging risks and hazards.

1.1.12 Documentation

The validation report should provide detailed documentation of validation testing results at all stages of new process development. The report should also include all elements of the validation test plan and a summary of the commercial verified process.

1.1.13 Role of process authority in the era of emergent novel and nonthermal preservation technologies

On July 2, 1971, the US FDA released a public warning after learning that a Westchester County, New York, man had died and his wife had become seriously ill from botulism after partly eating a can of Bon Vivant vichyssoise soup (1971 Bon Vivant botulism case).

The company began a recall of the 6444 cans of vichyssoise soup made in the same batch as the can that was confirmed to be contaminated. The US FDA then indicated that the company's processing practices made questionable for all products packed by the company and extended the recall to include all Bon Vivant soup—more than a million cans in all (https://ipfs.io/ipfs/QmXoypizjW3WknFiJnKLwHCnL72vedxjQkDDP1mXWo6uco/wiki/1971_Bon_Vivant_botulism_case.html#cite_note-Examination-3). The FDA ordered the shutdown of the company's Newark, New Jersey, processing plant on July 7, 1971. Ultimately, 5 cans of soup, out of 324, were found to be contaminated with botulinum toxin, all in the initial batch of vichyssoise soup that was recalled. Bon Vivant filed for bankruptcy and ultimately went out of business.

In August, 1971 canned chicken vegetable soup produced by the Campbell Soup Company tested positive for botulinum toxin. This finding resulted in the recall of some 4799 cans of soup from the implicated batch. This incident is reported to have cost Campbell's $5 million ($27.4 million in "today's dollars") and damaged the company's reputation.

These two public health incidents served to incentivize the US FDA and the National Canners Association to collaborate in promulgating the low-acid canned foods (LACF) regulation. A hallmark of these regulations was the creation of the processing authority.

"A processing authority is a person who has expert knowledge of thermal processing requirements for low-acid foods packaged in hermetically sealed containers, or has expert knowledge in the acidification and processing of acidified foods. Knowledge can be obtained by education or experience or both. Expert implies experience, knowledge and achievement as well as recognition as an authority on a subject, usually by one's peers. Anyone who is establishing scheduled processes must have adequate facilities for making the appropriate determinations (US FDA 21 CFR 113.83). Anyone who is evaluating processes which are less than the scheduled process (process deviations) must utilize procedures recognized by competent processing authorities as being adequate to detect any potential hazard to public health (US FDA 21 CFR 113.89)." It is notably that the regulations are intended to govern the end product of a processing authority's work rather than that person's qualifications. However, if CFSAN (HFS-617-Division of HACCP, Regulatory Food Processing and Technology Branch) is unfamiliar with the process authority (PA) they may review their qualifications, and the procedures and methods are utilized to evaluate the adequacy of the final work product.

In the United States, the role of the "PA" in the production of thermally processed, low-acid and acidified canned food is unambiguous. The concept of the PA has been enshrined in US FDA regulations since 1973. The situation in other regions of the world, however, is much less clear. Codex, for example, has in larger measure adopted and incorporated the US LACF regulations in its Code of Hygienic Practices for Low Acid and Acidified Low Acid Canned Foods (CAC/RCP 23-1979). The Codex standard does not have an explicit requirement for a PA. Instead the Code requires that scheduled processes for low-acid foods should originate from "Competent Persons" (CAC/RCP 23-1979—7.5.1.2). And while the nomenclature between FDA and Codex regulations are different, the activities and expectations of "competent persons" are implicitly the same as those of the LACF's PA.

The concept of the PA, in the United States, was expanded beyond low-acid and acidified canned foods by the California Almond Board and the USDA's Agriculture Marketing Service (2007) with establishment of the "California Almond Marketing Order." The Market Order requires the use of a "PA" for the validation of pasteurized or otherwise treated almonds to confirm a $4\log_{10}$ reduction of *Salmonella.* The Marketing Order states that the role of the "PA," in the context of its program, "is modeled after the FDA's LACF requirements." This action taken by a nongovernmental organization was a major development in the history of the PA in the United States.

1.1.14 Requirements for processing authorities [21 CFR113.83 and 113.89]

Ultimately a processing authority must establish thermal processes (US FDA 21 CFR 113.83). Without a properly established process, process control at the processor level is useless. In other words, you cannot verify that which has not previously been validated. Establishing processes requires among other things: (1) considerable knowledge about product characteristics and the effect of each processor's equipment and procedures on those characteristics of importance to HP; (2) experience in conducting HP, TD, thermal death time studies and other scientific methods; and (3) the ability to determine through evaluation of data generated by these studies and tests that sufficient testing has been accomplished to identify all possible *critical factors* that could affect the heating characteristics of the product and the safety of the final product. It is often necessary for a processing authority to recognize the inadequacies or inexperience of a processor in order to provide the processor with sufficient information to ensure understanding of what factors are critical for achieving food safety. These same responsibilities should apply whether the PA is establishing a thermal process or some other methods of food stabilization. These fundamental precepts of food safety science remain valid for nonthermal methods and for other food processing techniques and applications.

1.1.14.1 Novel and nonthermal processing technologies

US LACF regulations are unequivocal in that a PA is required for processors engaged in processing certain classes of high-risk, thermally processed, canned foods. While processing a food with a pH greater than 4.6, with a A_w exceeding 0.85, and the food is intended for nonrefrigerated distribution, a PA is required. By contrast, the regulations governing other classes of foods, processed by other than thermal methods, are less so emphatic. For example, a low-acid or acidified food, held in a hermetically sealed container, processed using high hydrostatic pressure (HPP), and intended for refrigerated distribution would not require, by regulation, the involvement of a PA. Another example for consideration might include an HPP-processed, low-acid, canned food (A_w less than 0.85) that is intended for ambient distribution. This food would not require a scheduled process and therefore no mandatory involvement of a PA. Another intriguing example is that of an aseptically packaged, acidified, low-acid food (pH < 4.6) treated by a PEF and also intended for ambient storage and distribution. This product and process combination would not necessarily be absolved of the requirement for a PA, even though PEF is considered a nonthermal process. This is a unique, if not novel, set of product and processing conditions, and therefore the processor would be advised to consult with a PA and also the FDA.

HPP in combination with heat, involving either an acidified or low-acid foods that are intended for ambient distribution must involve the expert services of a PA. PATS processes, for example, are regarded by the FDA as thermal processing techniques that require a scheduled process. The LACF regulations were modified in 2009, following the first-ever submission of a scheduled process for PATS, to accommodate this method of sterilization.

When the time and temperature components of a pressure-assisted process do not conform with those customarily used for thermal sterilization of low-acid food products, these processes too should require involvement of a PA. For example, the pressure-enhanced sterilization (PES) process (accepted by FDA in 2015) relies on relatively mild temperatures (110°C/230°F) and 700 MPa (for 3.0 minutes) to achieve commercial sterility in a low-acid, heterogeneous, food system. This process (while not representative of the classical 12 D thermal process) was, with the assistance of a PA, accepted by the FDA even though the agency noted, in its letter of no objection, that this was not a thermal process. A compelling argument could be made that PES is a pressure-driven nonthermal process, and as such, a process filing with the US LACF is not required. As of now this is a gray area in the LACF regulations. For the time being, processors using similar pressure-driven thermal processes are advised to work with a PA and to also file a scheduled process with the FDA.

1.1.14.1.1 Process adherence

Under the LACF regulation of the US FDA, sections 21 CFR 108.35(c)(3) and 108.25(c)(3) are written fundamentally for purposes of assuring that valid processes are used for mitigating the food safety risk that may be associated with thermally processed low-acid and acidified canned foods (https://www.fda.gov/low-acid-canned-food-manufacturers-part-1-adminstrative-proceduresscheduled-processes; https://www.fda.gov/low-acid-canned-food-manufacturers-part-2-processesprocedures).

These sections require all LACF and acidified food processors to comply with the mandatory requirements of 21 CFR 113 and 114, including registration of processing plants, filing of scheduled processes, control of critical factors, and adherence to the filed processes. Information concerning processes and other data provided to the agency are regarded as trade secrets. FDA's request and examination of such information do not constitute approval. However, when it is the agency's determination that the submitted data are insufficient or inconsistent with mitigating the public health risk, the FDA will reject the submitted process. Occasionally the agency will also issue a "*letter of no objection*" based on its evaluation of the submitted process information. Defining the process definition, elaborating the critical factors, and identifying the process control measures are the work of the PA. The benefits of the LACF regulations and the requirement for a PA are clear. The incidents of botulism in commercially canned food, since 1973, have plummeted to nearly zero.

The Hazard Analysis Risk Assessment and Preventive Controls (HARPC) rule of the Food Safety Modernization Act (2011) requires that CCPs of the company's food safety plan are overseen by a Preventive Control Qualified Individual (PCQI). While PCQIs are not process authorities, they do function to preserve process assurance for those foods that do not explicitly have a regulatory mandate for a PA. For example, HPP-processed apple juice (pH 4.5) intended for nonrefrigerated distribution would not have a statutory requirement for the involvement of a PA. In this case, the PCQI would be responsible for confirming that the applied HPP process resulted in the 5-$\log_{10}$ reduction of acid-adapted *E. coli* consistent with the requirements of the Juice HACCP Regulation (21 CFR 120). The distinction between the roles and responsibilities of the PCQI and PA is confusing in some circumstances and may lead to unqualified individuals assessing safety foods processed using novel methods.

A PA should be involved, at least, in the early research and design phases when novel preservation methods are considered. Their knowledge of microbiology, the processing equipment, and process control procedures will enable an assessment of regulatory requirements and also of the potential risk to public health.

1.2 Conclusions

It is important to recognize that each new process will be unique and the presented information can be considered only as recommendations. Each unique process suggests that validation studies are process and product specific. It is the responsibility of the food processors and PA to demonstrate the ability of their process to achieve the specific (required) log reduction of the pertinent microorganism in each portion of the product produced. In addition, process validation should lead consistently to high-quality and best sensory products with significant commercial benefits for the manufacturer and reduced risk of process failures.

Prospectively validating novel preservation methods is essential for protecting public health. Because of the newness of these processing techniques and myriad of unanswered questions, unsettled regulatory status, and also because of vagaries in the science pertaining to public health outcomes when using novel methods, it is advised that all perspective processors should engage with a PA for both technical and regulatory guidance.

References

1971 Bon Vivant botulism case Wikipedia. https://alchetron.com/1971-Bon-Vivant-botulism-case

Ahn, J., Balasubramaniam, V.M., Yousef, A.E., 2007. Inactivation kinetics of selected aerobic and anaerobic bacterial spores by pressure-assisted thermal processing. Int. J. Food Microbiol. 113, 321–329.

Borowski, A.G., Ingham, S.C., Ingham, B.H., 2009a. Lethality of home-style dehydrator processes against Escherichia coli O157:H7 and Salmonella serovars in the manufacture of ground-and-formed beef jerky and the potential for using a pathogen surrogate in process validation. J. Food Prot. 72, 2056–2064.

Borowski, A.G., Ingham, S.C., Ingham, B.H., 2009b. Validation of ground-and-formed beef jerky processes using commercial lactic acid bacteria starter cultures as pathogen surrogates. J. Food Prot. 72, 1234–1247.

CAC/RCP 23-1979.

Char, C.D., Guerrero, S.N., Alzamora, S.M., 2010. Mild thermal process combined with vanillin plus citral to help shorten the inactivation time for Listeria innocua in orange juice. Food Bioprocess. Technol. 3, 752–761.

Codex (CAC/GL 69-2008, Guidelines for the Validation of Food Safety Control measures).

Eblen, D.R., Annous, B.A., Sapers, G.M., 2005. Studies to select appropriate nonpathogenic surrogate Escherichia coli strains for potential use in place of Escherichia coli O157:H7 and Salmonella in pilot plant studies. J. Food Prot. 68, 282–291.

Fairchild, T.M., Foegeding, P.M., 1993. A proposed nonpathogenic biological indicator for thermal inactivation of Listeria monocytogenes. Appl. Environ. Microbiol. 59, 1247–1250.

Gurtler, J.B., Rivera, R.B., Zhang, H.Q., Geveke, D.J., 2010. Selection of surrogate bacteria in place of E. coli O157:H7 and Salmonella typhimurium for pulsed electric field treatment of orange juice. Int. J. Food Microbiol. 139, 1–8.

Hu, Gurtler, 2017. Selection of surrogate bacteria for use in food safety challenge studies: a review. J. Food Prot. 80, 9.

IFT, 2000. Kinetics of microbial inactivation for alternative food processing technologies. J. Food Sci. Suppl., Institute of Food Technologists. http://vm.cfsan.fda.gov/~comm/ift-pref.html.

Koutchma, T., Ho, G., Slade, P.J., 2005. Roadmap to validation of processing technologies for juices. Food Prot. Trends 2, 114–119.

Kuo, J., Chen, C., Nellor, M., 2003. Standardized collimated beam testing protocol for water wastewater ultraviolet disinfection. J. Env. Eng. 8, 773–779.

Lorentzen, G., Ytterstad, E., Olsen, R.L., Skjerdal, T., 2010. Thermal inactivation and growth potential of Listeria innocua in rehydrated salt-cured cod prepared for ready-to-eat products. Food Control. 21, 1121–1126.

Montville, T.J., Dengrove, R., de Siano, T., Bonnet, M., Schaffner, D.W., 2005. Thermal resistance of spores from virulent strains of Bacillus anthracis and potential surrogates. J. Food Prot. 68, 2362–2366.

Montville, T.J., de Siano, T., Nock, A., Padhi, S., Wade, D., 2006. Inhibition of *Bacillus anthracis* and potential surrogate bacilli growth from spore inocula by nisin and other antimicrobial peptides. J. Food Prot. 69, 2529–2533.

NACMCF, 2006. Requisite scientific parameters for establishing alternative methods of pasteurization. J. Food Prot. 69 (5), Supplement. NACMCF.

Niemira, B.A., 2010. Irradiation sensitivity of planktonic and biofilm-associated Listeria monocytogenes and *L. innocua* as influenced by temperature of biofilm formation. Food Bioprocess. Technol. 3, 257–264.

Orlowska, M., Koutchma, T., Kostryanska, M., Tang, J., 2014. Surrogate organisms for pathogenic O157:H7 and non-O157 *Escherichia coli* strains for apple juice treatments by UV-C light at three monochromatic wavelengths. Food Control. 47, 647–655.

Rouse, A.H., Atkins, C.D., 1955. Pectinesterase and pectin in commercial orange juice as determined by methods used at the Citrus Experiment Station. Bulletin of the University of Florida Agricultural Experiment Station, Lake Alfred, Florida, 570, pp. 1–19.

United States FDA, Title 21 Code of Federal Regulation 113.83 and 113.89.

United States FDA, 1987. Center for Drugs and Biologics and Center for Devices and Radiological Health. Guideline on general principles of process validation.

United States FDA, 2001. Hazard Analysis and Critical Control Point (HACCP): procedures for the safe and sanitary processing and importing of juice. Final rule. Federal Register, 66 (13). United States Food and Drug Administration, Washington, DC.

United States FDA, 2015a. FSMA final rule for preventive controls for human food. Docket Folder FDA-2011-N-0920. Retrieved from https://www.regulations.gov/document?D = FDA-2011-N-0920-1979.

United States FDA Federal Register, 2015b. Current good manufacturing practice hazard analysis and risk based preventive controls for human.

United States Food and Drug Administration, 2000. 21 CFR Part 179. Irradiation in the production, processing and handling of food. Federal Register 65, 71056–71058.

CHAPTER 2

Validation of high hydrostatic pressure process

Tatiana Koutchma[1] and Keith Warriner[2]
[1]Novel Food Processing Technologies, Agriculture and Agri-Food Canada, Guelph, ON, Canada
[2]Food Science Department, University of Guelph, Guelph, ON, Canada

2.1 History and introduction to high-pressure processing technology for foods

High hydrostatic pressure (HHP), ultrahigh-pressure processing, or high-pressure processing are encompassing terms that describes the utilization of elevated pressures, typically in the range of 100–800 MPa, with or without the addition of external heat. Although there are variations in the process, typical regime the product is vacuum packed and then pressurized by direct and indirect methods utilizing a pressure-transmitting medium (normally water). Pressure is applied homogenously (isostatically) throughout the chamber that means the high pressure treatments is independent of the pack geometry or size.

HHP of foods can be traced back to the late 19th century with the work of Hite who demonstrated the process could extend the shelf life of milk and fresh produce (Hite et al., 1914). Yet, with the advent of improved thermal processing technologies, the use of HHP for food preservation passed into history before being reconsidered as a nonthermal and mild preservation technique.

The incentive for employing HHP was a need to extend the shelf life of products while retaining the sensory characteristics of the product as an additional process to freezing, blanching, among other processes. Japan was the first to launch commercial HHP products that primarily included jams and sauces, then later for preserving avocado-based products. The advances of HHP have been rapid with concept to commercialization taking a matter of 2 years. The application of HHP as a pathogen-intervention step was only considered in the late 1990s with the advent of widespread introduction of Hazard Analysis and Critical Control Point (HACCP) (FDA; 2013). Implementation of HHP as a pathogen-

Validation of Food Preservation Processes based on Novel Technologies
DOI: https://doi.org/10.1016/B978-0-12-815888-3.00007-5

intervention step has been restricted due to the cost of pressure units that was in the order of $250,000 for a laboratory-scale equipment. As a consequence, there have been relatively few research groups from Europe and the United States working in the area of HHP, and certainly the knowledge base has lagged significantly behind technological advances that made pressure process commercially feasible. In effect this has led to a high demand from manufactures to apply HHP despite the lack of knowledge and a full understanding of the factors that affect the efficacy of the process or associated risks linked to specific food products. Indeed, there is no standard methodology for validating HHP or surrogates to verify performance.

From 2000 there was an increasing use of HHP in the United States as a pathogen reduction step to comply with regulations being introduced by the US Food and Drug Administration (FDA) and US Department of Agriculture (USDA) to enhance food safety of juices, deli meats among other products. Within this period, it was recognized that the pressure tolerance of pathogens was complex with many knowledge gaps, some of which still exist today.

2.2 Fundamentals of HHP

2.2.1 HHP principle of operation

The HHP operation starts with loading prepacked food products in perforated baskets, and the baskets are transported into the high-pressure vessel. The vessel is sealed and then filled with a pressure-transmitting fluid (normally water) and pressurized by the use of a high-pressure pump (an intensifier), which injects additional quantities of fluid. The packages of food, surrounded by the pressure-transmitting fluid, are subjected to the same pressure as that which exists in the vessel itself. After the product has been held for the desired process time at the target pressure, releasing the pressure-transmitting fluid decompresses the vessel. For most applications, products are held for 3–5 min at 600 MPa that allows approximately 6–8 cycles/h including time for compression, holding, decompression, loading, and unloading. Slightly higher cycle rates may be possible using fully automated loading and unloading systems and more powerful pumps and higher throughput intensifiers and/or more pumps and intensifiers. After HHP treatment, the processed product is removed from the vessel and stored or distributed in the conventional manner.

2.2.1.1 *Types of high-pressure processing*

2.2.1.1.1 High-pressure pasteurization

Pasteurization treatment typically employs pressures in the range of 200–600 MPa at or near ambient temperatures for a specific holding time (HT). HHP is typically considered a pasteurization step that can reduce the levels of pathogens and spoilage microbes and requires post-treatment refrigeration to minimize the outgrowth of survivors (Lakshmi et al., 2013).

2.2.1.1.2 Pressure-assisted thermal processing or pressure-assisted thermal sterilization

Sterilization of low-acid foods typically requires applying temperatures in the order of 121°C although this can be reduced to 90°C when heating is used in combination with high pressure within the range of 500 – 900 MPa. Inactivation of bacterial spores is the primary objective and the product is normally stored at ambient temperature (Sevenich et al., 2013).

2.2.2 HHP commercial applications

HHP application continues to increase around the world especially within Canada and the United States, which are home to more units than across the globe combined. The underlying reason for HHP being popular in North America could be attributed to the relatively straightforward regulatory approval (CFIA. 2017). It is also possible that the greater centralized manufacturing within North America provides the scale of economy to support HHP, and consumers are willing to pay a premium for the perceived greater quality. Within Europe there are a greater number of laboratory-based HHP units to support research although there is less utilization in commercial processing. The success of HHP implementation by a large number of companies demonstrates HHP is an effective food safety measure that can be applied to mitigate risk for a variety of foods and beverages.

Ready-to-eat (RTE) products are of particular concern for contamination with *Listeria monocytogenes* because they may support the growth of the pathogen during refrigerated storage and can be often consumed without further cooking. The examples of RTE products include deli products, hotdog, whole hams, sausages, meat salads, and other products that have been treated with a lethality step. Because the *Listeria* infectious dose is thought to be low, Food Safety Inspection Service (FSIS) has maintained a "zero tolerance" for the pathogen in RTE products. HHP treatment is recommended by FSIS as a postlethality treatment for RTEs.

Meat products, or more specifically RTE deli meats, comprise a significant sector in recent years. Within North America, HHP is a key for *Listeria* control and reduction of other pathogens. A number of processors use HHP in place of additives such as diacetate lactate to inhibit potential outgrowth although this may compromise shelf-life stability.

High pressure processing is primarily applied to treated fruit & vegetables such as avocado and apple based products. The main motivation for HHP of foods is to reduce pathogen and spoilage microflora, enzyme inactivation, gelation, and unfolding of allergenic proteins. Acidity and/or low water activity facilitates shelf-stability although low-acid starch products can be prone to premature spoilage by spore-forming bacteria such as *Bacillus*. The cold-pressed fruit and vegetable juice segment has been increasing and growing significantly in recent years.

The seafood industry has a relatively long history of using HHP for shell removal, but the process is considered a significant intervention to reduce *Vibrio vulnificus* which is a pathogen that can be associated with shellfish.

A commodity group underrepresented in the HHP sector is dairy products even though research has demonstrated the inactivation of *Listeria* in soft cheese products (Tomasula et al., 2014). Yet, the changes in texture due to pressure-induced denaturation and increased enzyme activity remain challenges to implementing the technology within the dairy sector (Chawla et al., 2011).

2.2.2.1 HHP-enabled product innovation

HHP can bring positive product attributes in terms of texture, biodigestibility and replacing the need for additives such as salt and other chemical preservatives. Consequently, it is possible to HHP in developing novel products for the consumer market.

Beverages processed using HHP help satisfy increased demand for juices and beverages rich in vitamins, functional compounds, and probiotics to strengthen the immune system. Numerous scientific reports support the fact that HHP does not change nutritional profiles, antioxidant content, or flavors. Plant-based dips and spreads produced using HHP are also consumer favorites .

Among grains and legumes, pea remains as an important protein source and one of the most sought-after ingredients for the development of meat alternatives. HHP demonstrated its potential as extended shelf life (ESL) treatment of pea protein based dips and spreads along with positive effects on product texture .

RTE vegan fitness meals such as a cereal porridge with mango puree and spinach—banana puree have been developed in Germany using HHP technology and resulted in 35 times longer shelf life than untreated product.

HHP treatment at 600 MPa and 3 min extended shelf life of Colucci pasta and pizza doughs for more than 3 months compared with only 3 days without treatment. The microbial counts of total aerobic bacteria, *Enterobacteriaceae* ., and other bacteria remained under the designated microbiological criteria for 8 months without the need of chemical preservatives.

Understanding of HHP as a pathogen inactivation step has been constrained due to the scalability issues, cost of labratory-scale pressure units, and availability for Containment Level 2 and 3 facilities.

The successful HHP implementation has led to a high demand from food manufactures to apply HHP despite the lack of a full understanding on factors that affect the efficacy of the process or associated risks linked to specific food products. Indeed, there is no standard methodology for validating HHP or established surrogates to verify performance. Yet, a scientifically sound validation process is a critical component in assessing the efficacy of HHP as a microbial-control measure.

This chapter will focus on HHP process validation used for nonthermal pasteurization that reduces pathogen levels and supports the reduction of spoilage vegetative microbes to extend shelf life of foods and beverages.

2.2.3 HHP microbiological effects

2.2.3.1 Mode of inactivation of microbes by HHP

The precise mode of inactivation of microbes by HHP remains to be fully elucidated although evidence to date indicates that a combination of cellular targets are affected (Li et al., 2013). The nature of cellular damage is dependent on the applied pressure. Specifically, at low pressure (50 MPa) the ribosomal function is inhibited. As the applied pressure is increased, protein denaturation is noted at 100 MPa with membrane damage occurring at 200 MPa. Up to 300 MPa, the membrane/protein damage induced by pressure is partially reversible, thereby enabling the cellular recovery. However, beyond 300 MPa, the damage is irreversible leading to cell death (Abe, 2007; Rastogi et al., 2007). The pressure-related mode of inactivation by HHP is relevant given that many laboratory-sized units are limited to lower pressures (<300 MPa) that could explain the contradictory results published in the literature. Yet, there is a consensus that the loss of cell membrane functionality and integrity are considered the key lethal effects caused by HHP. The evidence supporting the role of pressure on membranes is based on the loss of nutrient transport and energy-generating (ATPase) proteins/enzymes in pressure-treated cells (Tholozan et al., 2000). It is also noted that cells with saturated fatty acids making up membranes can be correlated to enhanced barotolerance compared to unsaturated fatty acids that are correspondingly sensitive (Bevilacqua et al., 2013).

Irreversible damage to enzymes is caused by the disruption of noncovalent bonds leading to denaturation and thereby metabolic function. Consequently, metabolic pathways halt leading to an oxidative burst of reactive oxygen species that ultimately results in the loss of cell viability (Simpson and Gilmour, 1997). Nucleic acids such as DNA are not degraded by HHP but become more compact although denaturation of polymerases results in growth inhibition (Mohamed et al., 2012).

2.2.3.2 HHP barotolerance

In general terms. it has been found that prokaryotes have higher barotolerance to pressure compared with eukaryotes with Gram positives exhibiting higher resistance than Gram negatives. Cocci have a tendency to have higher resistance compared to rods with the highest barotolerance being linked to endospores. Although still to be confirmed, thermophilic bacteria with increased saturated fatty acids within their membranes exhibit high tolerance to pressure than mesophiles or psychrophiles (Bartlett, 2002). In terms of specific bacteria, *Vibrio* has been reported to be most the pressure sensitive with a 4-log reduction being achieved at pressure in the order of 293 MPa for 120 s (Ma and Su, 2011). Yet, the relative resistance of different microbes is dependent on several factors that include strain selection, propagation methods, and intrinsic properties of the product matrix. Specifically, cells in the stationary phase exhibit higher pressure resistance to those undergoing exponential growth (Patterson, 2005). This makes sense given that growing cells have the need to express proteins and generate ATP. In contrast, stationary cells have low metabolic rates and frequently express stress proteins to further enhance pressure resistance (Robey et al., 2001).

It is widely accepted that the lethality of HHP is enhanced by low pH. However, less well investigated is the effect of food components on the enhancing or reducing barotolerance. For example, protein, sugars, and lipids can provide a protective effect towards high pressure. Milk proteins can enhance the barotolerance of *Staphylococcus aureus* compared to when simple buffer solutions are applied or suspended in meat slurries (Smiddy et al., 2004).

One of the most reported aspects of HHP but little understood is the significant variation in barotolerance of different species, strains, and even subpopulations within a monoculture. From studies performed to date it has been reported that acquisition of pressure resistant is not universal. For example, it has been demonstrated that from a range of bacteria which includes *Yersinia, Salmonella, Shigella, Pseudomonas, Listeria*, and *Escherichia coli*, only the latter acquired barotolerance (Vanlint et al., 2013). Even then, not all *E. coli* exhibited pressure resistance and was limited to two strains. Of the strains tested, *E. coli* MF1655 adapted to pressure is the most barotolerant vegetative cell encountered with 800 MPa being required to reduce populations by 2.8-log colony-forming unit (cfu) (Vanlint et al., 2013). *E. coli* O157:H7 can also attain pressure resistance after being exposed to sublethal pressures. The barotolerance of *E. coli* could persist over sequential generations in the absence of pressure although gradually lost. The underlying mechanisms of pressure resistance remain unclear although considered to be through stress protein expression and reduced cell volume (Vanlint et al., 2013).

Grouping of microbial types with respect to pressure tolerance is difficult to determine given the influence on strain type, supporting matrix, processing conditions, and cultivation conditions to recover survivors. Yet, through comparisons of inactivation data the general trend is that bacterial endospores exhibit the highest barotolerance with yeasts being the most sensitive. Vegetative bacterial cells have intermediate pressure resistance although exceptions exist. For example, *Vibrio* is commonly found to be pressure sensitive relative to other bacterial types (Table 2.1). *E. coli* can exhibit high barotolerance although strongly strain dependent with no differentiation between pathogenic and nonpathogenic types. Yet, there is a tendency of Gram positives being more resistant compared to Gram negatives although this depends on the food matrix, in addition to processing parameters (Patterson, 2005). Molds have intermediate pressure tolerance being inactivated between 300 and 600 MPa; although given foods are vacuum packed, the risk of outgrowth is minimal (Smelt et al., 2002).

Table 2.1 *D* and *Z* values for the high hydrostatic pressure inactivation of different microbes.

Microbe	P_{ref} (MPa)	T_{ref} (°C)	Log D_p	Z_p (MPa)	Z_T (°C)
Bacillus spores	400	100	0.27	614.7	45.2
Bacillus anthracis	400	NA	2.33	876.4	NA
Clostridium spp.	400	100	0.85	616.3	20.4
Clostridium sporogenes	400	100	0.70	1088.6	67.2
Cronobacter	400	NA	0.19	368.2	NA
Escherichia coli	400	30	0.88	385.6	97.5
Listeria spp.	400	30	0.56	298.9	38.6
Vibrio spp.	400	30	0.058	206.9	18.4
Zygosaccharomyces bailii	400	30	0.84	91	141.7

D_p, Decimal reduction time; Z_p, change in pressure to D_p by 1-log unit; Z_t, change in temperature to change D_p by 1-log unit; *NA*, not available.
Source: Modified from Farakos, S.M.S., Zwietering, M.H., 2011. Data analysis of the inactivation of foodborne microorganisms under high hydrostatic pressure to establish global kinetic parameters and influencing factors. J. Food Protect. 74 (12), 2097–2106.

Applied pressure is one parameter that defines the inactivation kinetics of microbes by HHP. In general, growing cells are more pressure sensitive to those in the stationary phase due to the expression of stress responses in the latter (Norton and Sun, 2008). However, HHP temperature is one of the key parameters that affect inactivation kinetics. The highest barotolerance within microbes of interest exists between 20°C and 40°C with increased efficacy of HHP at colder or higher temperatures. The only exception is observed with *Lactococcus lactis* that increases in barotolerance with temperature up to the point of thermal inactivation (Sonoike et al., 1992). The underlying mechanisms for the relation of temperature and pressure on microbial inactivation are considered to be due to the increased compressibility of water and cell cytoplasm with decreasing temperature, thereby enhancing the transmission of pressure (mechanical) energy (Lori et al., 2007).

2.2.3.3 Matrix effects on barotolerance of bacterial cells

The matrix in which the bacterial cells are suspended in during HHP has a significant influence on the observed barotolerance. For example, low water activity provides protection to microbes against HHP (Setikaite et al., 2009). Moreover, survival of bacterial cells under high pressure is affected by the type of solute used (i.e., sugar or salt) (Considine et al., 2007). Specifically, sucrose provides a greater protective effect to cells exposed to high pressure compared to NaCl (Setikaite et al., 2009).

Acidity also contributes to enhancing the efficacy of HHP treatment, thereby enabling lower pressure to be applied to achieve the target log 5 reduction. For example, *L. monocytogenes* can be decreased by an additional 3-log cfu by HHP when suspended in a buffer at pH 4 compared to pH 6. The pH effect on sensitizing cells to pressure is not only dependent on the degree of acidity but also on the acid type. Here, it is theorized that pressure treatment results in a shift in the associated acids relative to dissociated forms. The associated (protonated) form of the acid can freely permeate the cell membrane leading to acidification of the cytoplasm that also impacts the ability of the bacterium to revive during posttreatment storage.

The presence of antimicrobials has also been reported to sensitize bacteria to high-pressure treatment. Membrane acting antimicrobial agents such a bacteriocins, essential oils and lytic enzymes (lysozyme) act synergistically to enhance the lethality of high pressure against microbes. It has been reported that essential oils exhibiting synergistic activity with HHP include allyl isothiocynate, tertiary butylhydroquinone (TBHQ), carvacrol, and citral that enable >5-log cfu reduction of bacterial pathogens at pressures in the order of 100–300 MPa (Gayan et al., 2012).

The sensitivity of *Salmonella* to HHP can be enhanced in the presence of nisin through a mechanism, which is thought to be through destabilization of ribosomes. Here, in the presence of nisin, the pressure required to support an 8-log cfu reduction of *Salmonella* was reduced from 500 MPa to 350–400 MPa depending on the strain tested (Lee and Kaletunc, 2010). Similar synergistic effects of HHP with nisin have also been observed with *E. coli* O157:H7 and *L. monocytogenes* (Alba et al., 2013). Lysozyme has also been demonstrated to sensitize bacterial cells to high pressure through enhancing the activity of the lytic enzyme (Hauben et al., 1996).

2.2.3.4 Pretreatment culture conditions on barotolerance

The pressure tolerance of bacteria is dependent on the both the intrinsic resistance of the cell and imposed pretreatment stress. For example, growing cells are more sensitive to high pressure compared to growing cultures (Patterson, 2005). This makes sense given that stationary cells have down-regulated metabolism and expressed stress proteins to protect cell structures (Robey et al., 2001). The imposition of osmotic or cold stress also induces stress proteins that contribute to barotolerance (Hormann et al.,

2006; Lado and Yousef, 2002; Wemekamp-Kamphuis et al., 2004). This is relevant given that bacterial cells are frequently subjected to stress prior to high-pressure treatment although this fact is not always considered when determining inactivation kinetics.

It has been speculated that pressure resistance is encoded by specific genes beyond that of stress protein expression (Huang et al., 2014). Supporting evidence for barotolerance genes is that certain strains of *E. coli* exhibit enhanced pressure resistance and can be selected by sublethal pressure cycles (Vanlint et al., 2013). In one example, *E. coli* MF1655 could be adapted to pressure and resist treatments up to 800 MPa (Vanlint et al., 2013). *E. coli* O157:H7 can also attain pressure resistance after being exposed to sublethal pressures. The barotolerance of *E. coli* could persist over sequential generations in the absence of pressure although this trait was lost through subculturing. Interestingly, from screening the genomes of resistant *E. coli* strains illustrated that the pressure resistance was accompanied by the loss of nutrient transporters, aerobic metabolism, fitness factors, and mobile genetic elements (Garcia-Hernandez, 2015). Therefore it appears that genes that promote high growth rates and substrate utilization counter the resistance to pressure. It has also been reported that *E. coli* strains harboring the island locus of heat resistance that encodes for stress proteins enhance pressure resistance (Li et al., 2013). This has led to the thought that overexpression of heat shock proteins in mutants leads to increased pressure resistance (Aertsen et al., 2004). The same finding has been observed for *Saccharomyces* (Palhano et al., 2004) and *Lactobacillus rhamnosus* (Ananta and Knorr, 2004), where the expression of heat shock proteins correlate to pressure resistance. The most diverse barotolerance has been observed in *L. monocytogenes* with threefold differences in pressure resistance between parent and resistant strains. Through studies it was reported that pressure-resistant strains of *L. monocytogenes* had a broader tolerance to thermal, oxidative, and acid stresses (Karatzas and Bennik, 2002). It was subsequently reported that a common feature of barotolerant *L. monocytogenes* was a deletion in a gene regulator that essentially would cause overexpression of stress genes within the bacterium at the cost of virulence (Karatzas et al., 2003). Collectively, it would appear that pressure resistance is most notable in dormant cells or in those with overexpression of stress proteins to stabilize cellular structures.

There is significant intra-strain variability in relation to bartolerance and consequently it has been challenging to identify index strains that can be applied in verification trials of HHP processes. This is compounded by the lack of standard protocols to undertake validation trials that makes comparisons between studies challenging in terms of bartolerance assessment of individual strains. Therefore attempts have been made to correlate the resistance of different stresses to that of barotolerance. There were initial attempts to correlate thermal resistance to pressure tolerance, although observed in certain cases was not universal. Interestingly, a negative correlation has been reported for toxigenic *E. coli* whereby strains that were inhibited by relatively low salt concentrations (<2%) exhibited increased resistance. Although not studied in detail, it was proposed that the early expression of stress proteins led to increased barotolerance. The approach using osmotolerance to predict pressure resistance could be applied as a predictor for barotolerance. Yet, given that stresses applied prior to treatment influence pressure resistance, it is unlikely that no single metric could be applied for predicting tolerance.

2.2.3.5 Posttreatment storage and recovery

The observed log reduction of bacteria is dependent on the post-storage treatment and methods used to recover survivors. HHP essentially dismantles the ability of a cell to function with a decrease in culturability over time. Consequently, if survivors are enumerated immediately following HHP, the log count reduction (LCR) is lower compared to when the analysis is performed 72 h post-treatment. In a similar manner, survivors enumerated on nonselective agar recover higher counts compared to when selective media is applied. For example, by using nonselective agar the recovery of *Salmonella* was increased compared to selective XLT4 agar (Zhou et al., 2020).

The extent of post-treatment recovery of *L. monocytogenes*, *Salmonella*, and *E. coli* O157:H7 was independent of headspace composition (anaerobic vs aerobic) but was enhanced at low temperatures (<4°C) compared to room temperature (Juck et al., 2012). Although the result would suggest holding samples under temperature abuse conditions would enhance the apparent lethality of HHP treatment, it should be noted that this will also promote the growth of surviving pathogens like *Salmonella* (Koseki and Yamamoto, 2006; Toledo et al., 2012). Low acid also promotes the potential recovery of bacteria treated with high pressure although increasing the acidity promotes inactivation.

2.2.4 HHP quality effects

It is well established that the primary structure of protein remains unaffected, hence the protein quality of HHP-treated meat/poultry/other high-protein food is expected to remain unchanged. Yet, there is little work performed on the positive or negative effects of high pressure processing on the digestibility and functionality of proteins (Toro-Funes et al., 2014).

Water-soluble vitamins (e.g., vitamin C, vitamins B_1, B_2, B_6, and folic acid) are generally said to be unaffected by HHP. However, information in the scientific literature suggests variable effects/retention of vitamins in different food products depending on the treatment conditions. Studies that address the mechanisms of vitamin disappearance and pressure-induced changes in profile with respect to nutritional significance would be useful in evaluating the nutritional safety of HHP treatment. Similar principles would also apply for fat-soluble vitamins.

The effect of HHP on metal ions/speciation of metals like iron and the effects on nutritional quality of HHP-treated foods should be explored.

2.2.5 HHP effects on allergens

There has been a significant increase in the risk posed by allergens and remains the leading cause of product recalls. The World Allergy Organization estimates that 220–250 million people suffer from food allergies across the globe (Demoly et al., 2014). The leading allergens of concern include peanut, bovine milk, seafood, soybean, wheat, tree nuts, and egg. A common feature of allergens is the high stability to acids and heat (Sicherer and Sampson, 2010). The effect of HHP processing on allergens has been found to be dependent on the type and processing parameters (Table 2.2).

The mode by which HHP inactivates microbes has yet to be fully elucidated but is through to occur through a combination of direct and indirect effects. The direct mechanism is through pressure-induced denaturation or conformational changes that alter the structure of the protein to the extent that it fails to illicit immune response. Indirect mechanism includes the extraction of allergen through dissociation of plant cell structure or through enhanced enzymatic hydrolysis. Importantly, there has been no evidence that HHP generates allergenic proteins through reactivity of food components.

2.2.6 HHP effects on patulin mycotoxin

Mycotoxins are secondary metabolites produced by a certain molds and when ingested can result in acute or chronic conditions such as cancer. There are diverse ranges of mycotoxins encountered in foods with patulin

Table 2.2 Effect of HHP processing on selected allergens encountered in foods (Huang et al., 2014).

Allergen	HHP conditions	Comments	Reference
Peanut	150–800 MPa, 20°C–80°C, 10 min	No change in allergen structure	Johnson et al. (2010)
Soybean	300 MPa, 40°C, 15 min	Decrease allergen content of soy sprouts	Penas et al. (2011)
Whey protein isolate	200–600 MPa, 30°C–68°C, 10–30 min	Increase allerginicity of lactoglobulin	Kleber et al. (2007)
Apple	400–800 MPa, 80°C, 10 min	Decreased allergen content	Johnson et al. (2010)
Shrimp	450 MPa, 40°C, 60 min	No change in allergen structure	Shriver and Yang (2011)

being associated with fruit juices such as apple (Moake et al., 2005). It has been reported that HHP treatment in the range of 300 MPa, 500 MPa, and 800 MPa at 20°C for 60 min led to the reduction of original patulin content (155 mg/kg) by 42%, 53%, and 62% in apple juice respectively (Bruna et al., 1999). Avsaroglu et al. (2015) reported that the patulin concentration in clear apple juice was decreased from 100 ppb to 76 ppb (24% reduction) by applying a high pressure treatment of 300 MPa for 5 min.

In a comparative study, the HHP-assisted degradation of patulin in different apple-based juices [pineapple:apple:mint, apple:carrot:beet:lemon: ginger and romaine:celery:cucumber:apple:spinach:kale, parsley:lemon (GJ] was assessed (Hao et al., 2016). The extent of patulin degradation could be significantly correlated with the sulfhydryl group concentration of the juice with ascorbic acid and nitrite being less significant. HHP treatment of 600 MPa for 300 s at 11°C resulted in a 62 ppb decrease in patulin introduced into GJ juice which also contained the highest level of thiols (97 μM). The thiol concentration of the other juices ranged between 39 and 69 μM with a corresponding decrease in patulin of 43–49 ppb following the same HHP treatment. The study has illustrated that HHP can be applied as a risk management tool to control patulin in apple-based beverages although the extent of mycotoxin degradation is dependent on processing conditions and composition of the juice.

In addition to patulin, deoxynivalenol and zearalenone have also been demonstrated to be highly sensitive to degradation by HHP. It was reported that a treatment of 550 MPa for 20 min at 45°C could completely degrade mycotoxins on grains (Woldemariam and Emire, 2019).

2.2.7 HHP effects on enzymes

One of the main incentives for applying HHP processing as a pasteurization step is retaining enzyme activity to maintain the "natural" or "raw" characteristics of products, such as vegetable juices. However, enzymes may also reduce the shelf life of products by supporting degradation reactions, promoting browning and generating off-odors.

Enzymes respond differently to HHP with some being denatured while the activity of others is enhanced. The majority of studies to date have focused on the spoilage enzymes, such as polyphenol oxidase, pectin methylesterase, peroxidase, and lipoxygenase (Table 2.3). The effect of HHP on enzymes is strongly dependent on the food matrix with activity been decrease, unchanged or increased. The increase in activity can be attributed to reorientation of the enzyme or simply the release from lysed cells.

2.3 HHP critical product and process parameters

2.3.1 Product parameters

HHP process variables may be further complicated by the presence of variability of food composition (e.g., fat content and its distribution); pH and water activity; and physical product parameters, such as density, heat capacity, and thermal conductivity.

2.3.1.1 pH

Inactivation of vegetative pathogenic and spoilage microorganisms in high-acid ($pH < 3.5$), medium-acid ($3.5 < pH < 4.6$), and low-acid foods ($pH > 4.6$) has been studied (Pilavtepe-Celik, 2013). Reduced pH is generally synergistic with pressure in eliminating microorganisms.

Pressurization of food constituents causes the ionic dissociation of water molecules with a corresponding decrease in pH (Hoover et al., 1989). The dissociation in aqueous solutions is influenced by temperature and pressure leading to different reversible pH changes of pressurized buffer systems. The effect is more pronounced for phosphate buffers than water-based food systems. On the other hand, Tris-HCl-type buffers are more stable under pressure (Heinz and Knorr, 2002). As pH is lowered, most microbes become more susceptible to HHP inactivation, and the recovery of sublethally injured cells is reduced. Accordingly, the type and composition of the buffer

Table 2.3 Effect of HHP processing on enzymes in different fruit and vegetable matrices.

Enzyme	Source	HHP conditions	Response	Reference
Pectin methylesterase	Apple	200–600 MPa, 15°C–65°C, 0.5–10 min	Increased activity with pressure up to 40°C	Baron et al. (2006)
	Carrot	100–600 MPa, 25°C, 10 min	40% inactivation at 100 MPa28% increase in activity at 600 MPa	Kim et al. (2001)
	Orange	500–900 MPa, 25°C	10%–93% inactivation	Goodner et al (1998)
	Tomato	400–800 MPa, 25°C	50% increased activity at 400 MPa	Shook et al. (2001)
Polyphenol oxidase	Apple	Up to 900 MPa	Inactivation >600 MPa	Weemaes et al. (1998)
	Avocado	689 MPa, 20 min	78% inactivation	Palou et al. (1997)
	Carrot	100–600 MPa, 25°C, 10 min	90% inactivation 500 MPa	Kim et al. (2001)
Peroxidase	Apple	600 MPa, 20°C, 15 min	Twofold increase	Prestamo et al. (2001)
	Carrot	500 MPa, 20°C, 1 min	70% increase	Anese et al. (1995)
	Orange	400 MPa 32°C, 15 min	50% inactivation	Cano et al. (1997)
Lipoxygenase	Avocado	689 MPa, 20 min	95% inactivation	Palou et al. (1997)
	Carrot	300 MPa, 10 min, 25°C	83% inactivation	Kim et al. (2001)
	Tomato	300 MPa, 20°C, 10 min	60% inactivation	Van Buggenhout et al. (2006)

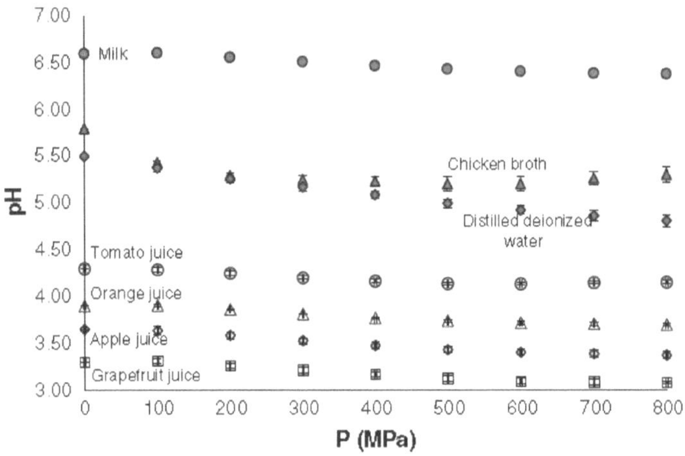

Figure 2.1 pH (25°C) of the low-viscosity liquid foods and distilled deionized water under high pressure measured by the pH sensor. *Modified from Samaranayake, C.P., Sastry, S.K., 2013. In-situ pH measurement of selected liquid foods under high pressure. Innov. Food Sci. Emerg. Technol. 17, 22–26.*

used in the microbial testing should be documented along with microbial inactivation data (Samaranayake and Sastry, 2013). Foods subjected to *in situ* measurements included: grapefruit juice pH of 3.30; ranch dressing 3.35; apple juice 3.65; orange juice (unsweetened) 3.90; guacamole 3.90; yogurt (plain nonfat) 4.28; tomato juice 4.30; distilled deionized water 5.50; chicken broth (99% fat free) 5.80; and milk 6.60 (Samaranayake and Sastry, 2013). The measurements showed an increase of acidity in all tested liquid food and water samples with increasing pressure (Fig. 2.1).

In general, food samples exhibit a buffering action under pressure resisting changes in pH. Unlike the pH of distilled deionized water, the pH of fruit juices and milk remained almost unchanged up to 100 MPa, and then gradually dropped and leveled off beyond 500 MPa (Fig. 2.1). The maximum pH drop was, however, 0.3 pH units in the entire pressure range (0.1–800 MPa). For chicken broth (Fig. 2.1), a drastic pH drop (-0.37 ± 0.02 pH units) ($P \leq .05$) was observed with increasing pressure up to 100 MPa, and then the pH further dropped to an average of 0.05 pH units per 100 MPa and bottomed out at 500 MPa. Increase in pH beyond 500 MPa was, however, insignificant ($P > .05$), averaging about 0.03 pH units per 100 MPa. As these foods are known to contain weak acid buffering agents, the observed buffering action under HHP is therefore not surprising. The semiliquid foods (Fig. 2.1) also underwent a pH drop (in the range of -0.3 to -0.4 pH units) ($P \leq .05$) at 100 MPa. For ranch dressings and yogurt, the pH again dropped significantly ($P \leq .05$) at 200 MPa, and then bottomed out between 300 and 600 MPa, whereas for guacamole, the pH was virtually unchanged for the rest of the pressure range (100–800 MPa). The maximum pH drop (-0.58 ± 0.05 pH units) was observed with yogurt in the pressure range of 300–600 MPa. The results with guacamole (Fig. 2.1) further indicate that even for semiliquid samples, the US National Institute of Standards and Technology (NIST) pressure–density data for water can be used for pH determination using the model. The authors related the drastic pH drop in the chicken broth and semiliquid foods at 100 MPa to the large compressibility of these fat- and air-containing foods (Min et al., 2010), resulting in pronounced hydrogen bonding that improves the movement of protons. It was also found that the pressure-dependent pH change is apparently independent of the concentration of added antimicrobial acids. Thus any potential strategy of increasing the level of antimicrobial acid addition for the enhancement of process would not likely be effective.

2.3.1.2 Water activity

Studying the effect of water activity (*Aw*) on the HHP inactivation of *Listeria innocua*, Hayman et al. (2007) found that reducing the water activity reduces the inactivation rate observed by an increase in *D*-values. The same effect of water activity on barotolerance has been observed from a broad spectrum of microbes including bacteria and yeast (Palou et al., 1997; Ueno et al., 2011).

2.3.1.3 Fat or oil content of the food

During pressure treatment, the temperature of different food components increases due to physical compression and returns back to its initial value upon decompression. The heat of compression of most of the high-moisture food materials is very similar to that of water, that is, 3°C per 100 MPa at 25°C. However, fatty foods have increased compression heating, relative to water, due to higher compressibility with long-chain unsaturated fatty acids and lower specific heat.

Food matrix may provide additional protective or inhibitory effects. Compared to orange and tomato juices, milk showed a considerable baro-protective effect against HHP inactivation of *Salmonella enterica* and *L. monocytogenes* (Xu et al., 2009). The inactivation patterns of *S. enterica* and *L. monocytogenes* varied in different food matrices. Decimal reduction times

Table 2.4 Summary of effects of product parameters on HHP pasteurization.

Product parameter	Range investigated	Effect on lethality of HHP	Controlled by product formulation	Research studies published
pH	3–7	Yes	Yes	Yes
A_w	0.9–1.0	Yes	No	Yes/No
Food Matrix		Yes	Yes	Limited
Fat content	Depends on product type	Yes	No	Limited

(*D* value) for *S. enterica* treated at 300 MPa were 0.63, 0.94, 0.41, and 0.45 min, respectively, in deionized water, milk, orange juice, and tomato juice, while those for *L. monocytogenes* were 1.40, 9.56, 1.11, and 0.94 min, respectively (Xu et al., 2009).

The effects of product matrices on HHP inactivation of microbes are summarized in Table 2.4.

2.3.1.4 Physical properties of foods under high pressure

Food pressurization results in changes in the thermodynamic properties, the rheological properties, and compression heating that may impact safety and quality. Knowledge of pressure dependency on various thermal and physical properties, such as thermal conductivity, specific heat, density, of food materials is needed for evaluating heat transfer within processed volume. While pressure affects primarily the volume of the system, the temperature can cause both volume and energy changes in the system.

After the pioneering work of Bridgman (1931), the properties of water under pressure were well documented. Data are available from the International Association for the Properties of Water and Steam (IAPWS). A software implementation of IAPWS work can be obtained from the NIST. Most high-moisture foods are assumed to have thermodynamic properties similar to those of water at the pressures and temperatures used in HHP. This assumption is less valid for lower moisture, high solids, and nonuniform systems. Very limited information is available on properties of food materials under pressure because of the practical challenges associated with the in situ measurement of these properties at elevated pressures (Ramaswamy et al., 2005).

2.3.1.5 Compressibility

Compressibility is an intrinsic property of the material and is defined by the balance between attractive and repulsive potentials. Compression of a liquid decreases the average intermolecular distance and tends to reduce rotational and translational motion. Food material (e.g., orange juice) is considered to contain molecules that occupy space in excess of that needed for close packing. This excess is called "free volume," and it is this volume that is reduced in initial compression (Rasanayagam et al., 2003). At elevated pressures, when the free volume has largely disappeared, reduction in Van der Waals dimensions may occur and the compressibility is greatly diminished. Isothermal compressibility (β) is defined as the relative change in volume (V) with pressure (P):

$$\beta = -\frac{1}{V}\left(\frac{dV}{dP}\right)_P \tag{2.1}$$

Compressibility and density were reported by Min et al. (2010) for 16 foods based on volumetric measurements at 25°C and pressures up to 700 MPa. This included sucrose solutions (2.5%–50%), soy protein solutions (2.5%–10%), soybean oil, chicken fat, clarified butter, chicken breast, ham, cheddar cheese, carrot, guacamole, apple juice, and honey. Compressibility decreased with pressure at a rate that decreased with increasing pressure. For equal mass concentrations, protein solutions were less compressible ($P < .04$) than sucrose solutions at pressures greater than 200 MPa. Relative to water, fats showed high compressibility to 100 MPa, similar compressibility from 100 to 300 MPa, and less compressibility from 300 to 700 MPa. Chicken breast, ham, cheddar cheese, carrot, and guacamole showed relatively large compressibility from 0.1 to 100 MPa. Honey showed the smallest volume decrease over 700 MPa. Variability in compressibility of different materials is likely due to differences in concentration, chemical composition, and complex interactions between components within a food system.

Thermal expansion of materials is caused by temperature change during compression. Thermal expansion coefficient (α) is another thermodynamic property which provide a measure of the amount by which the density changes in response to a change in temperature at constant pressure (P):

$$\alpha = \frac{1}{V}\left(\frac{dV}{dT}\right)_T \tag{2.2}$$

Estimation of volume [Eq. (2.1)] or thermal expansion [Eq. (2.2)] under pressure was constrained by practical challenges associated with developing reliable sensors/instrumentation that can withstand elevated pressure conditions.

2.3.1.6 Thermal conductivity

Very limited studies reported thermal conductivity (k) of food materials under pressure. Denys and Hendrickx (1999) studied k of tomato paste and apple pulp at pressures up to 400 MPa. Zhu et al. (2007) studied k values of potato and cheddar cheese up to 350 MPa at 5°C. Material thermal conductivity increased with increase in pressure. It was influenced by the amount of moisture present in the food material. Ramaswamy and Balasubramaniam (2007) reported thermal conductivity of selected liquid foods under pressure. Water and water-like substances (apple juice) were found to have the highest k values (up to 0.82 W/m°C at 700 MPa and 25°C), while fatty foods such as canola oil and clarified butter have the lowest (0.29–0.4 W/m°C, respectively, at 700 MPa and 25°C) values. Honey and high-fructose corn syrup had intermediate values. The pressure dependency of thermal conductivity values was higher at lower moisture contents than at higher moisture contents. This may be due to the smaller percentage of incompressible water and higher porosity of the sample at lower moisture contents. The estimated k values of all the food materials tested under pressure were higher than corresponding k values of materials under atmospheric pressure signifying faster thermal equilibrium under pressure (Ramaswamy and Balasubramaniam, 2007).

2.3.1.7 Specific heat

The data on specific heat of pure water as a function of elevated pressure and temperature are readily available through the NIST database. These values were approximately 10% lower than those estimated at atmospheric pressures. In the absence of experimental data, researchers often ignore the effect of pressure on specific heat in heat transfer calculations.

2.3.1.8 Density

When food materials are processed under pressure, significant change in the volume of the product occur possibly due to reduction in the "free volume" between molecules or the compactness of voids occupied by air. The density of materials under pressure can be estimated by knowing the change in volume of materials under pressure and its mass. Density is then calculated as the ratio of mass to volume. The volume change of materials under pressure can be estimated by using linear velocity differential transducer (Bridgman, 1931). Min et al. (2010) reported density increase of 16 food samples at a decreasing rate. Density under pressure for foods with relatively high solids content, fat content, and porosity deviate from the behavior of water, which is often used as an input in temperature distribution simulations.

2.3.2 Process parameters

Typical temperature and pressure cycles during high-pressure sterilization [high pressure—high temperature (HP-HT, solid line)] and pasteurization [high pressure—low temperature (HP-LT, dotted line)] cycles are schematically shown in Fig. 2.2. HHP is characterized by three critical parameters: time, pressure, and temperature as shown in the plots.

2.3.2.1 Time

Cycle time. The total time for loading, closing the vessel, compression, holding, and decompression and unloading is commonly referred to as the "cycle time."

Pressure come-up time (CUT). The time required to increase the pressure of the sample from atmospheric pressure to the target process pressure. The CUT depends upon the rate of compression of the sample and pressure-transmitting fluid and is proportional to the power of the pump and flow rate, size of intensifier (gallons per minute (gpm)), restrictions of high-pressure tubing, valve loss and applied pressure.

Decompression time or come-down time (CDT). The time to bring a food sample from process pressure to near atmospheric pressure. The CDT is almost instantaneous. The fast cooling capacity of HHP is of most interest in the production of high-quality foods. The effect of CUT on pathogen inactivation by HHP has not been studied to any great extent although is known to vary within the range of 0.5—3.8 min depending on pressure unit being used (Haiqiang, 2007; Kingsley et al., 2007; Ramaswamy et al., 2003). It can be considered that similar to thermal processing, there will be a degree of lethality of HHP as the pressure ascends to the pressure set point. The effects of changing rates of compression and decompression on inactivation of *S. aureus* during HHP were reported by Syed et al. (2014) in skim milk, orange juice, and Tris buffer samples. The lethality with slow raise time was considered to be due to the ability of the bacteria to induce stress proteins, thereby providing protection against HHP. For rapid compression times, the

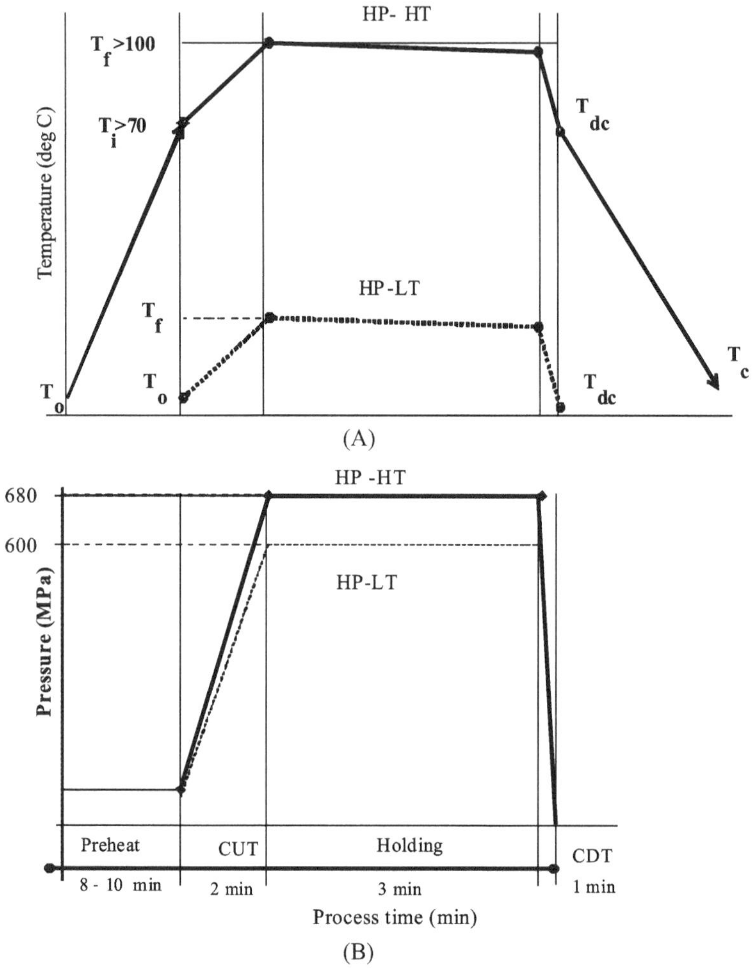

Figure 2.2 Temperature and pressure histories during high-pressure sterilization (HP-HT, solid line) and pasteurization (HP-LT, dotted line) processes. (A) temperature histories during pasteurization and sterilization; (B) pressure histories during pasteurization and sterilization process. Symbols represent product temperatures at initial (T_o), after preheat (T_i), during HP (T_f), after decompression (T_{dc}), and after cooling (T_c) processes.

cell has insufficient time to express stress proteins, thereby more susceptible to pressure. Earlier, Vogel et al. (1999) reported that the process of rapid pressurization (6.7 MPa/s) and slow decompression (0.83 MPa/s) was more effective than the process of same maximum pressure and HT but reverse pressurization and depressurization rates. Noma et al. (2004) reported that rapid decompression rates (<2 s) were more effective at inactivating *E. coli* O157:H7 compared to slow decompression (30 s).

Pressure holding time. The time interval between the end of compression and the beginning of decompression.

Commercial processing times, which is the sum of CUT + CDT + HT, can range from a pulse of a few seconds to over 20 min. Short-processing times must be targeted since process time has a significant effect on the commercial economics of HHP (Balasubramaniam et al., 2004). A processing time less than 5 min is preferred to maximize productivity and economically justify use of the technology. Obviously, long CUTs will add appreciably to the total processing time and affect the process throughput, but these periods will also affect the inactivation kinetics of microorganisms. Therefore consistency and control of the CUT, HT, and CDT are important in the development of HHP techniques.

Process pressure (P) refers to the holding pressure during the sample treatment. It should be noted that although a set pressure point is defined for a process, the actual applied pressure oscillates around a set point. Therefore although a set point of 600 MPa is applied, there will be times when the pressure could fall to 595 MPa or lower. The ability to maintain the set point is a function of the sensors to monitor the pressure and hydrolytic rams to respond. It has been recommended to control and record pressure at F 0.5% (electronic) or F 1.0% (dial display) level of accuracy. Further, it is recommended to have at least two methods to measure pressure and have an appropriate periodic calibration program. Frequently, a reference sensor or gauge is sequestered from routine operation and only used in a periodic check mode. Since most HHP users do not have calibration equipment in-house for high pressure, experienced laboratories should perform the calibration of any reference device.

Pressure pulsing. Refers to treatments whereby two or more pressure pulses are applied. The parameters to consider include the duration and frequency of pulses along with peak pressure and processing temperature.

Fig. 2.2 shows schematically the examples of HHP temperature and pressure cycles during HP-LT and HP-HT processing utilized for pasteurization and sterilization, respectively. The CUT, the CDT, and the duration of the processing cycle shown in the figure are specific to the design of HP vessel used.

2.3.2.2 Temperature

Temperature has been identified as a major factor that affects the efficacy of HHP along with pH and the magnitude of the applied pressure (Farakos and Zwietering, 2011). With respect to HHP, the effect of temperature relates to that applied during processing, in addition to changes occurring during the course of a pressure run.

Product initial temperature (T_i): The initial temperature of the product before the commencement of pressurization.

Product process temperature (T_f): The final product temperature during pressurization.

Product final temperature: The product temperature after decompression (T_{dc}) and cooling (T_C).

Process temperature: Temperature of pressure-transmitting fluid (water) in the vessel during HT.

Heat of compression: The instantaneous temperature change in materials during compression or decompression is often called the heat of compression.

Fluids such as air, water, and food materials undergo adiabatic heating during compression, and cooling during decompression. The following Eq. (2.3) expresses this adiabatic heating

$$\frac{dT}{dP} = \frac{T\alpha_p}{\rho C_p} \tag{2.3}$$

As indicated by Eq. (2.1), the temperature increase dT depends on the volumetric expansion coefficient α_p (measured in units of K^{-1}), the density ρ (kg/m^3), and the isobaric heat capacity C_p (J/kg/K) of the materials well as on its initial temperature T (K).

In general, the temperature of both the product and the pressure-transmitting fluid may rise by 20°C–40°C during HHP treatment. Solid metallic materials do not experience significant compression heating. Therefore the temperature increase may vary in foods with relatively complex compositions (Patazca et al., 2007).

The quasiadiabatic temperature increase δ_S upon compression of food samples can be evaluated as the ratio of the temperature increase to the pressure increase using the following Eq. (2.4):

$$\delta_s = 100 \times \frac{\Delta T}{\Delta P} = 100 \times \frac{T_f - T_i}{P - P_{atm}} \cong 100 \times \frac{T_f - T_i}{P}\left[\frac{°C}{100\ MPa}\right] \tag{2.4}$$

where T_f is the temperature of the sample at the applied pressure, T_i is the initial temperature, and P is the applied pressure. Water, carbohydrates, fats, and proteins are the main components of the complex food matrix that respond uniquely under compression. It has been reported that water has the lowest rate of temperature increase under compression,

Table 2.5 Compression heating (°C/100 MPa) of selected food substances determined at initial product temperature of 25°C.

Food substances	Temperature increase (°C/100 MPa)
Orange juice, tomato salsa, skim milk, salmon fish	2.6–3.0
Carbohydrates	2.6–3.6
Proteins	2.7–3.3
Mayonnaise	5.0–7.2
Extracted beef fat	6.2–9.1
Olive oil	6.3–8.7

Source: Data from Balasubramaniam, V.M., Ting, E.Y., Stewart, C.M., Robbins, J.A., 2004. Recommended laboratory practices for conducting high-pressure microbial inactivation experiments. Innov. Food Sci. Emerg. Technol. 5 (3), 299–306.

about 3°C per 100 MPa at 25°C, whereas fats have the highest value, up to 6.7°C–8°C per 100 MPa (Patazca et al., 2007). Nevertheless, only limited information is available on the temperature rise with compression of real foods with complex fat compositions (Table 2.5).

As followed from Eq. (2.1), under compression heating conditions, the sample temperature at process pressure is dictated by the sample's thermodynamic properties. In reality, heat transfer between the sample and the pressure-transmitting fluid or the pressure vessel during the CUT and hold time will likely influence the measured sample temperature.

Preheating, precooling, and temperature equilibration of products are important steps in HHP to achieve the required process temperature. A uniform initial target temperature T_i of the food sample is desirable in order to achieve a uniform temperature increase in a homogeneous food system during compression. If cold spots are present within the food or the food system is not homogeneous, parts of the product will not achieve the target process temperature T_f during pressurization. In other cases, keeping the initial temperature low (~4°C) can assist in avoiding a temperature increase during pressurization, and a precooling step may be required.

Initial temperature of pressure-transmitting fluid and the process vessel must be documented, as they are integral conditions for microbial inactivation. Similar to a thermal process, sufficient equilibrium time (primarily based on dimensions and thermal properties of the test product sample and vessel) should be given to assure that all locations within the pressure vessel and the product are equilibrated at ±0.5°C of the target initial temperature. For heterogeneous food samples, additional care or time may be needed to achieve equilibration. An adequate description of methods used for preheating or controlling the sample temperature before HHP should be provided.

The temperature control includes control of the temperature of the pressurization water from 5°C to 30°C. However, there is no control of the water temperature inside the vessel. The high-pressure vessel does not have any temperature control, and it is at the temperature of the room where it is located.

It has been recognized that processing high pressures of interest do not influence the type K thermocouple readings at temperatures below 500° C. The reference thermocouple sensor should be located at the cold point or equivalent zone within the pressure vessel and calibrated to an accuracy of F 0.5 degrees (Metrick et al., 1989). Standard methods and good laboratory practices regarding temperature measurement should be followed. It is important to consider that all compressible substances change the temperature during physical compression and this is an unavoidable thermodynamic effect.

Detailed information on CUT, HT, CDT, medium pressure and temperature, product pressure and temperature (before, during, or after HHP) are usually not provided in the published studies. Sometimes the pressure increasing and/or decreasing rates are given. The effects of HHP process parameters as reported in reviewed literature are summarized in Table 2.6.

2.4 Packaging

2.4.1 Requirements and types of packages

Packaging is a vital factor for any preservation method to be successful since it protects the food products from recontamination during processing and adverse environmental conditions during storage. HHP is a batch process in which food materials are packaged before the application of high pressure and the food product and the package are both exposed to the processing conditions so the entire pack remains a "secure unit" until the consumer opens it.

Prepackaging helps to prevent contamination from the pressure medium and improves HHP efficiency. However, any damage or alteration in the package or packaging materials can potentially cause the loss of its hermetical properties. Therefore quality, safety, and shelf life of the food product can be adversely affected.

Table 2.6 Summary of effects of process parameters on the efficiency of HHP pasteurization.

HHP process parameter	Range	Effect on lethality reported	Capability to control parameter in processing	Research performed
Come-up time, min CUT	2–3	Yes	No	Limited
Come-down time, min CD	0.3–0.4	Yes	No	Yes
Holding time, min HD	3–5	Yes	Yes	
Process time CUT + HT + CD, min	≥ 5	Yes	Yes	Yes
Process pressure, MPa	400–600	Yes	Yes and no	Yes
Initial temperature, °C	4–20	No	Yes	Reports of initial temperature no effect
Process temperature, °C	20–50	Yes	Yes	Available
Product temperature, °C	4–30	No	No	No

In-container processing requires packages in the form of pouches, large bulk bags, or container–lid combinations such as bottles, tubs, or trays. 90% of HHP foods are expected to be processed in flexible or partially rigid packages. Similar to food products, packaging materials also undergo compression heating. Knoerzer et al. (2010) have shown that polypropylene and polyethylene undergo compression heating greater than water under both HP and low and mild temperatures of 10oC and 50oC and HHP-high temperature (90oC) conditions up to 750 MPa. In particular, the temperature increase with pressure is not linear and, therefore, the relative increase with respect to water depended on the pressure range selected as well as the initial temperature. Since in typical industrial process 5% of HP vessel is filled with packaging, compression heating behavior of packaging materials may affect heat transfer by forming insulation layer surrounding food and consequently temperature distribution within HP vessel.

2.4.1.1 Requirements for plastic packaging materials

1. Flexibility: A packaging material needs to be flexible enough to withstand compression forces so that it can avoid irreversible deformations and maintain its physical integrity. Typically, high-barrier flexible pouches made of polymers or copolymers with at least one flexible side can be used for processing solid or liquid food products.
2. Package design: Package size and shape are also critical in terms of maximizing the number of packages, which can be fitted in the chamber. Therefore proper package design can contribute to the economical processing.
3. Maintaining integrity: Packaging materials used for HHP need to endure harsh treatment conditions so that they can sustain the physical integrity both during the process and extended storage. Maintaining the integrity during both the process and storage is particularly important in terms of the safety and quality of the food products. Ideally, as well as maintaining its mechanical and other built-in features, packaging material needs to protect consumer acceptance and convenience (Lambert et al., 2000). Barrier, mechanical, and mass transfer properties (sorption and migration) of the package must be resistant to the changes occurring during the process. During the process, temperature increases and the volume of the package decreases due to the pressure applied. Once the pressure is released, the package needs to regain its original conditions.
4. Barrier properties: Packaging material needs to have desired barrier properties to maintain the quality of food products especially during the shelf life. Water and oxygen are two of the most important components in food which determine the quality and shelf life. Barrier properties of the polymers are very vital considerations for shelf life, and type of polymer chosen plays a crucial role at this point. Developing packaging materials with improved water, oxygen, and light impermeability properties is essential to ensure desired shelf life for high pressure and pressure-assisted thermally processed food products.
5. Sealing is an important point for flexible pouches, and seal strength should be maintained during the processing in order to avoid any leakage.

Minimum headspace: Vacuum packaging is important for a uniformity of HHP treatment because air in the package has higher compressibility. Excessive amount of gases can increase the CUT of the processing. Vacuum packaging can also help avoiding oxygen-related reactions including lipid oxidation during both processing and storage. Vacuum packaging can improve loading factor, and more packaged product can be processed at one time in the limited volume of the pressure vessel.

2.4.2 Food contact compliance

Manufacturers have the responsibility to ensure there is no migration of chemical components from the packaging to the food. A variety of international bodies maintain standards on this. Some of these bodies such as the Canadian Food Inspection Agency (https://food-nutrition.canada.ca/food-safety/referencelist/index-en.php) and US FDA maintain a database for acceptable HHP materials.

Current research suggests that for many standard packages chemical migration is not significantly affected, and in some cases can even be reduced. However, relevant documentation should be sourced from the packaging supplier to ensure that the material is suitable for HHP with your product.

2.4.3 Packaging validation

Packaging should be obtained from reputable suppliers who can provide technical support, including specifications for package sealing parameters.

Packaging must be stored in a way that prevents contamination. It should be wrapped and stored off the floor until used in production.

Before a package is used, it should be tested for quality assurance with the actual food product. In cases such as low-viscosity juices, many products can be simulated by water. The processors should consider the following to test packaging:

- Create a worst-case scenario of packaged product to gather evidence to ensure the package selected for the product is actually will withstand established HHP conditions,
- Exaggerate headspaces,
- Overwhip product to incorporate air,
- Test the bottom and the top range of sealing parameters and the effect over several packages,

- Use colder product temperatures to test condensation and film brittleness,
- Challenge product at a higher pressure,
- Increase the process time,
- Run the cycle up to three times. As a processor, not all cycles will be 100% complete and rework may be required. By testing multiple cycles, assurance can be gained that reworked packages will not fail after HHP.

Also, after sealing and HHP treatment, a proportion of the products should be visually examined on both external faces for evidence of an incorrect seal or damage to the seal area. Packaged product showing any signs of defects should be rejected. A suitable method should be selected for testing the seals of the packs.

The following tests may assist in testing for packaging acceptance: bloat tests, dye tests, seal tests, drop and burst tests, etc. These quality checks should be carried out on randomly selected production packs after the primary and secondary sealers as well after HHP treatment.

Product must be labeled in accordance with the corresponding legal requirements.

2.5 HHP commercial systems

2.5.1 Design

The primary components of an HHP system include a pressure vessel (or shell), a closure or closures for sealing the vessel, a device for holding the closure(s) in place while the vessel is under pressure (e.g., a yoke), low-pressure fill systems to fill the pressure chamber with a temperature-regulated pressure medium (water) and remove air when the system is closed, a high-pressure intensifier pump or pumps to build up the pressure using a food-grade hydraulic oil, a system for controlling and monitoring the pressure and (optionally) temperature, and a product-handling system for transferring product to and from the pressure vessel. The machinery required is complex and requires extremely high precision in its construction, use, and maintenance.

2.5.2 Pressure vessels

Autofrettage vessels, wire-wound vessels, and heat shrink vessels are most frequently built for systems with pressures higher than 400 MPa.

Autofrettage (French, "self-shrinking" or autoshrinking) is a special mechanical process resulting in the pressure vessel prestressed condition after the autofrettage process completion. The process begins by pressurizing the shell that is stopped when the mechanical stress distribution through the shell wall is such that the internal wall is deformed plastically and the external wall is deformed elastically. After the pressure is released, the elastically deformed outside shell has the tendency to regain its original shape but is thereby hindered by the plastically deformed inside part. This creates a condition when the outer part of the shell pushes continuously onto the inside part resulting in the shell permanently under prestressed condition. When a shell is constructed using autofrettage, it can resist much higher internal pressures, comparing to the identical shell size made of a simple monoblock design. Autofrettage also makes shell much more resistant to crack growth.

Heat shrink technique is developed to induce beneficial residual hoop stresses in the cylindrical shell. In heat shrink the shell needs at least two separate layers; however, it can be made of three or more separate layers, assembled together. In the food processing industries typically, the inner layer is made of stainless steel, high-strength steel alloy inner layer, and high-strength steel alloy outer layer. Shrink fitting is accomplished by heating up the outer layer and after achieving the thermal expansion the cool layer (with earlier inserted stainless steel liner) is assembled. Finally, the whole assembly is allowed to cool down to room temperature, causing residual hoop stresses. The use of preexisting residual stresses in the cylinders before operation is essential to allow very high pressures by enhancing the load-carrying capacity and increasing the number of cycles of the operation. Heat shrink increases pressure vessel lifetime, durability, maximum pressure, and reduces the weight of shell; researchers succeeded in designing multilayer heat shrink fitted cylinders.

Wire-winding technique combines a single-layer cylinder and a continuous length of wire wound around it, at the prestressed condition. It is usually applied to increase the strength to weight ratio and to improve the fatigue life. The required wire-winding force is obtained based on constant effective stress in all layers at the maximum applied force. However, the failure of the wire (breaking or corrosion) will cause pressure vessel failure. The principle used in the wire-wound vessels is that they will leak before they break.

In some cases, the cylinders and the vessel frame are prestressed by winding wire under tension layer upon layer. The tension of the wire compresses the cylinders, reducing the diameter of the cylinders. This special arrangement allows an equipment lifetime of over 100,000 cycles at pressures of 680 MPa or higher.

Stainless steel construction is used in key areas of the vessel for reliability and easy washing down in the most demanding environments. The corrosion protection meets regulatory requirements for food processing plants. The systems also have provisions for filtering and reusing the compression fluid (usually water or a food-grade solution). As pressure vessels of all types utilize potentially hazardous energy, the relevant regulations (ASME Boiler and Pressure Vessel Code, Section VIII Division 3) seek to identify good design, good manufacturing practices (GMPs), and detailed safety assessments for the safe operation and maintenance of the vessels and auxiliary parts.

2.5.3 Commercial HHP manufacturers

The HHP equipment market is estimated at $350 million and is projected to grow at 11% compound annual growth rate (CAGR) for the next 5 years. Batch and semicontinuous systems are currently commercially available as options for HHP equipment. Semicontinuous HHP systems are used in only a few cases, to directly process pump liquid products and beverages which then need to be aseptically packaged.

Traditionally, HHP has been characterized by processing prepacked products using the so-called in-pack process. Although there are variations in the process, in the typical approach, a product is vacuum packed and then pressurized by direct and indirect methods utilizing a pressure-transmitting medium (normally water). The pressure is isostatic and thereby acts homogenously throughout the chamber and product independent of product size, shape, or dimensions.

The HHP operation starts with loading prepacked food products in perforated baskets, which are transported into the vessel. Then the vessel is sealed and filled with water that is pressurized by the use of intensifiers that inject additional quantities of water, consequently compressing the water by almost 15%. After the product has been held for the desired time at the process pressure, releasing the pressure-transmitting fluid decompresses the vessel. For most applications, products are held for 3–5 min at 600 MPa, which allows for approximately 6–8 cycles/h, including time for compression, holding, decompression, loading, and unloading.

Two key improvements—increased vessel size and reduced cycle time—allow for the intensification of batch HHP production capacity. Considering HHP equipment that can operate at 600 MPa, in 1998 the largest industrial vessel had a volume of 215 L. In 2009, the vessel capacity doubled and reached 420 L, and in 2014 the largest vessel ever designed by HHP equipment company Hyperbaric with a capacity of 525 L became commercially available. However, due to the size and shape of the bottles, vessel filling efficiency is often restricted to 40%–55%. To optimize HHP machine filling efficiency, it is important to reduce the amount of air inside the packaging, preferably using optimally designed vacuum packaging. With a smart approach, it is possible to create a honeycomb structure inside the vessel and fill all the space in the basket with products. Vacuum packages and pouches normally achieve filling efficiency of 50%–60%. This value is reduced when it comes to bottles and increased when it comes to process pouches.

2.5.3.1 Semicontinuous HHP for beverages

HHP has commercial applications for a broad range of product categories, including beverages, meat, poultry, seafood, dairy, fruit, vegetables, and RTE meals. The use of HHP for beverages and juices is the fastest-growing segment with 25%–30% market share. Processing beverages in-bulk in a semicontinuous system before bottling without any limitations for use of glass or metal packaging is of interest to the industry as an alternative to in-pack configuration. The early semicontinuous HHP machines were built as single vertical vessels (up to 210 L) or medium or small vessels (up to 3 × 20 L) connected in a series with a free or floating piston that could move up and down within the vessel. After the vessel was filled with a product to 85%–90%, the pressurizing fluid was pumped and compressed into the vessel on the opposite side of the free piston, displacing it and consequently compressing the product. However, because of corrosion and sanitation issues caused by direct contact of the product with the piston, vessel, plugs, and seals, semicontinuous HHP systems had limited use and no commercial success.

Avure Technologies (USA-Sweden), Hyperbaric (Spain-USA), MULTIVAC (USA-Germany), and BaoTau KeFa (China) are major suppliers of commercial-scale pressure equipment. Both horizontal and vertical pressure-vessel configurations are available. The horizontal-cylinder design is intended to serve in installations, where vertical space is limited and minimal plant alteration is desired. The horizontal orientation allows single-direction inline product flow. In addition, there are differences in capacities, cycle times, achievable pressures, capital costs, etc.

HHP equipment owners also offer toll services to food industries. This business model allows any food or beverage manufacturer to have access to industrial HHP equipment without the need of directly investing capital in the purchase of HHP unit. Instead, food manufacturers pay for the use of existing HHP industrial plants for their service on a toll basis.

AVURE Technologies offers vertically and horizontally oriented commercial pressure vessels. The typical HHP system blocks include pressure vessel, basket load conveyor, operator's panel, control system, water system, and high-pressure pumping systems. Avure patented QUINTUS wire-winding technology to contain isostatic pressure in the system. Pressure vessels and frames are prestressed and wound completely with high-tensile cold-rolled spring steel ribbon. Prestressing causes the forged steel vessel wall to remain in residual compression, even when subjected to maximum loads. This results in extremely high resistance to fatigue, eliminates tensile loads, and prevents crack propagation and brittle failure. Another feature is that wire-wound vessels will leak before they burst. The range of pressure vessel volume of Avure horizontal machines varies from 100 to 525 L for a maximum pressure of 600 MPa at maximum filling capacity of 70%.

The Avure 525L-600 measures up to the production needs of today's high-volume food processor, offering the highest throughput in the industry with features to minimize operating costs and maximize uptime. Improved filling and pressurizing efficiencies result in higher throughput, larger volume, and 10 cycles/h with a 3-min hold time. The larger diameter also accommodates the processing of larger and odd-shaped products.

HIPERBARIC (Spain-USA) designs, manufactures, and markets HHP equipment since 1999. Hyperbaric HHP single vessels' volume ranges from 55 to 525 L.

The 525 L is the largest size single HP vessel offered by Hyperbaric. The projection is that with the throughput of 3000 kg/h, this unit can offer the least expensive cost of HHP-processed product.

The Wave 6000/300 Tandem system (maximum working pressure of 600 MPa) from Hyperbaric is the biggest tandem HHP system for industrial production. In this system, two vessels and their peripherals work together, sharing the same intensifier pumping group. The operation is dephased: when one vessel is in the phase of increasing pressure, the other one is in the holding-time phase. Owing to its volume and vessel diameter (two vessels, each vessel of capacity 300 L and 300 mm in diameter), it is the most productive machine in the range.

2.5.3.2 Continuous "In-bulk" HHP

Hiperbaric recently introduced an innovative "bulk" approach that avoids the use of a free piston and the direct product contact with the vessel and plugs and delivers filling efficiency of at least 90%. The continuous HHP system contains a large, watertight flexible plastic bag or container located in the vessel where the liquid is processed. The bag is filled through a clean pressure-resistant valve allocated in the high-pressure plug. Depending on required productivity, the system can have one or two vessels of 525 L together with a system of tanks to store a beverage before and after pressurization.

The HHP pressurization cycle of the in-bulk process is similar to a conventional in-pack cycle. However, prior to liquid pressurization, three operations must be carried out: (1) clean-in-place of prerinsed complete circuit with water followed by the circulation of a cleaning agent at a high temperature and final rinse step with hot water; (2) steam sterilization of pipes and outlet tank; (3) mounting a bag in the vessel and attaching it to the hygienic pressure valve in the plug through which the beverage will be filled. By overcoming in-pack processing limitations, an in-bulk system reduces processing and packaging steps from six to four, increases filling efficiency from 45% to 55% up to 90%, and increases productivity from 3000 L/h to 5000 L/h. Although the cost of a Hiperbaric 525 Bulk machine is 20% higher than a similar in-pack unit, advantages of the in-bulk process include reduction in operational cost, labor, and energy consumption, and added packaging versatility. French processor Ateliers Hermes Boissons is the first company to put Hiperbaric's 535 Bulk machine to use commercially.

Hyperbaric also offers integration of its HHP systems into any existing production environment:

- Automation for product handling, loading, and unloading.
- Ancillary equipment for the HHP units, and peripherals such as hoppers, product customized HHP containers and basket for loading and unloading systems, post-HHP product dryers, etc.

UHDE HIGH PRESSURE TECHNOLOGY (UHPT), an HHP technology provider for more than 90 years, established a partnership with Multivac expertise in high-quality packaging machines to serve the food market with specialized HHP packaging solutions. Since 2010, UHPT has manufactured batch HHP machines with a portfolio of 2–350 L and special solutions. UHPT built the first pressure-assisted thermal sterilization plants up to a size of 150 L for 700 MPa and a temperature range up to 135°C.

MULTIVAC HHP (Germany) offers industrial single and tandem systems with vessel volumes from 55 up to 350 L. The maximum working pressure is 600 MPa, and cycle time (including 3-min pressure time) is from 7.0 to 9.0 min. Maximum treated volume of product is 1900 L/h and typical filling degree of 60%–80%. The tandem process HHP 700 increases process output up to 5000 L/h due to cycle time decrease to 3–5 min. MULTIVAC HHP is suitable for handling vacuum packs as well as modified atmosphere packs, for which MULTIVAC has developed a unique procedure.

BAO TOU KEFA High Pressure Technology Company from China offers industrial and laboratory-scale HHP units. The range of vessel volumes varies from 30 to 200 L at a maximum pressure of 600 MPa. The units are built using horizontal layout with the installation size: 5000 (length) × 2500 (width) × 1600 (height) (mm). The water is used as pressure medium, and working temperature can achieve 80°C. The company offers single, quos (2 × 200 L and 2 × 300 L) and quartos systems (4 × 200 L and 4 × 300 L) to increase production output. The company employs double heat shrink sleeve and autofrettage technology that assures maximum pressure vessel safety at lower weights, thus reducing facility costs.

Therefore four main HHP equipment manufacturers can be identified around the world in the United States, Spain, Germany, and China. The vertical and horizontal orientations of HHP vessels are the main designs currently used. The volume of commercial HHP vessels varies from 55 to 525 L with a filing capacity of 60%–80% that can run approximately 100,000 cycles at 600 MPa.

For comparison reasons, it should be noted that the evaluation of cycle time or total cycle time differs among manufacturers that consequently may affect the output evaluation. Hyperbaric includes loading and unloading times along with CUT, CD, and HT into the total cycle time. Avure excludes HT from the total cycle, whereas Multivac includes the HT into the cycle calculation.

2.5.4 Scalability of pilot and commercial-scale HHP processes

There are a very limited number of studies that investigated and reported scalability issues in HHP process development. The majority of the published reports used laboratory or pilot-scale machines. There is a lack of equipment description and technical details. Usually only the brand name and the model are provided.

Nine manufacturers of HHP equipment were identified that make and offer laboratory/pilot-scale machines. The range of process pressure varies from 600 up to 981 MPa with a maximum reported pressure of 1400 MPa (Resato HHP, The Netherlands). The range of process temperature varies from −20°C up to 110°C allowing developing processes in combination with high or low temperatures. Only a few machines are equipped with thermal probes. However, from machines description it is difficult to conclude whether the temperature of treated product can be measured and controlled under pressure conditions. The volume size of the vessels varies in the range of 0.03−3. 5 L for laboratory units and 35 L for pilot-scale units meaning that majority of food products cannot be HHP treated in their final commercial packages and corresponding mass and volume. Usually the food samples need to be repacked in the smaller size plastic bags in order to be processed in laboratory/pilot-scale units.

Also, laboratory and pilot systems without some external temperature control of the vessel will lose or gain heat faster than commercial systems during HHP, due to the larger surface area-to-volume ratio. This means it is important to know the expected temperature profile in the commercial equipment. If smaller HHP equipment is used for a validation study, the time−temperature profile should be comparable. The variation between HHP treatments using laboratory/pilot- or commercial-scale units can derive based on the following differences:

- Pressure-transmitting medium
  - o Water is used in commercial machines
  - o Water, water−glycol mixtures, and other fluids are used in laboratory-scale machines
- Temperature of pressure-transmitting medium
- Rate of pressurization and depressurization
- Drop of pressure during HT
- Food package weight, size, and shape
  - o Smaller repackaged products are used in the pilot studies
- Sample temperature
- Cooling methods and cooling rate

During product/process development trials, the effort should be made to use commercial product formulation, packaging, and HHP cycle as close as possible to the attributes of commercial product and process parameters. The objectives of the pilot-scale studies should be clearly outlined.

2.5.5 Calibration and maintenance of HHP systems

All monitoring devices need to be periodically calibrated including pressure/time/temperature measuring systems and records must be kept. Calibration procedures have to be traceable to national standards.

The pressure measuring instruments should be regularly calibrated. A two-pressure sensor set up with mutual checking can be used without requiring a manual check. The recommended level of accuracy to control or record pressure is $\pm 0.5\%$ (electronic) or $\pm 1\%$ (dial display). A master pressure gauge, if used for checking vessel sensors, should be calibrated annually.

Temperature-monitoring devices used for the process should be calibrated to the recommended accuracy level of $\pm 0.5\%$. Equipment used to control the time of the process should be calibrated to an accuracy of $\pm 1\%$.

HPHP units must be fitted with a control system that is capable of generating permanent process records that enable you to identify process deviations.

HHP systems have an automated control of the process during pressurization, holding pressure, and decompression. Keeping processing records and control product movement is required. The storage of electronic or manual records should be maintained with the relevant standards.

Also, system leaks can be detected by monitoring the time to pressure or the frequency of the HP pump use when maintaining target pressure. Detection of a leak should result in the machine being immediately stopped to prevent the HP pump being overloaded and minimize the risk of wear to seals and damage to components.

Equipment should be installed and commissioned by skilled people.

To ensure that the HHP unit is working properly, maintenance of the unit is crucial. It is recommended that manufacturer's instructions are followed for maintenance to extend the operating life of the system and reduce chances of breakdowns. Preventative maintenance is always recommended because the stress to other parts of the equipment when there is a gradual failure or breakdown could lead to more costly downtime.

Seals on HHP machines will require specific maintenance and need to be cleaned and exchanged on a regular basis. Correct training is important for the HHP equipment end users and cleaning teams to ensure seals are not damaged when changed or being cleaned.

2.6 International regulatory requirements

The introduction of HHP products in the 1990s was primarily related to extending shelf life in high-value foods with less emphasis being used for pathogen reduction. However, the introduction of pathogen control programs (i.e., HACCP) stimulated interest in HHP as a pathogen reduction intervention. Acceptance of HHP as a pathogen reduction step was problematic due to the large knowledge gaps that existed with respect to microbial inactivation. This led to different approaches by regulatory agencies in approving HHP products. Specifically, the main division in regulatory approaches was if HHP products should be considered as Novel Foods (Health Canada, 2010).

The appropriate definition of a "Novel Food," depends on the global location. Some countries have adopted specific term(s) and definitions (and regulations), whereas others have not. There are six definitions of Novel Foods in the European Union (EU), Great Britain, Canada, Australia, New Zealand, and China (New South Wales Food Authority, 2016). There are no regulations or formal definitions for "Novel Food" in the United States. GRAS (Generally Recognized As Safe) regulations can serve as some analogy of Novel Foods in the United States. Novel Foods and Ingredients are regulated in a varying manner by country, with the majority of systems based on a risk or safety assessment review model, with most also requiring notification and approval.

An alternative to the "Novel Foods" is the application of HHP as a pathogen reduction step in overall HACCP-based control program. Here, there is only a need to demonstrate a designated level of pathogen reduction as opposed to meeting extensive list of requirements for Novel Food approval.

2.6.1 Australia and New Zealand

Food Standards Australia New Zealand (FSANZ) development system is a cooperative arrangement between Australia, New Zealand, and the Australian States and mainland territories. Food safety programs based on HACCP principles are required for high-risk businesses.

FSANZ defines a Novel Food as those products that are not considered traditional and/or have been processed using a non-traditional technology. Foods designated novel are required to undergo a public health and risk assessment prior to being approved for sale. Categories of novel foods may include foods produced from new sources or by a process not previously applied to food. According to the record formed by the FSANZ Novel Foods Reference Group, HHP foods that do not require a prescribed pasteurization step, or where HHP is used in addition to pasteurization are viewed as traditional, not novel food. This justified by the application of HHP is a well-established food processing technology. The use of HHP for traditional foods that do not have a prescribed pasteurization step does not make a food nontraditional. (An example is HHP-treated yogurt.)

The FSANZ has issued guidelines on the application of HHP as an intervention step. Similar to the United States, New Zealand has adopted an outcome-based approach whereby an HHP must achieve a designated log reduction of the most relevant vegetative pathogen. It is the responsibility of the processor to identify the main pathogen of concern in a product along with what constitutes an acceptable pathogen reduction. From FSANZ guidelines provided, a 2-log reduction in pathogen levels would be deemed acceptable in some cases, although they are reviewed on a case-by-case basis (Food Standards Australia New Zealand, 2004).

The guide also takes regulatory elements from the EU with respect to the HHP process achieving the microbiological criteria for a designated product (e.g., the absence of *L. monocytogenes* in RTE deli meats). A further requirement stipulated in the New Zealand guide is that HHP products should have additional hurdles to prevent the outgrowth of bacterial endospores within the expected shelf life of the product and storage conditions. Other requirements are in line with the EU and the United States with respect to HHP unit specifications in terms of monitoring facilities (pressure and temperature), in addition to safety standards, as designated by Pressure Equipment Directive.

To circumvent the use of HHP to render unacceptable raw materials compliant, the New Zealand guideline stipulates that ingredients for preparing products intended for HHP must be accompanied by appropriate certificate of analysis. In addition, raw materials should be treated and stored in accordance with current GMPs. This part of the guide is intended to prevent HHP from being used to compensate for poor handling practices.

The guidelines for acceptance of HHP products also put forward the criteria for undertaking validation trials that are more defined compared to those published by other regulatory bodies. Yet, with the large knowledge gaps that exist with regards to HHP, the recommendations for validation studies are vague and primarily left to the operator to justify. An interesting feature of the recommendations is the acceptance of previously published results and modeling to justify the HHP conditions. This is a departure from the "Novel Foods" standards where a case-by-case evaluation is required. In the case when challenge trials are performed, the LCR achieved is reflective of the expected levels to be encountered under normal operating conditions, in contrast to the 5-log reduction expected in a process in the United States when prevalence data and scientific justifications are not available. A further feature of validation trials is the inclusion of shelf-life studies that must go beyond the expected time period proposed by the operator to account for the potential of posttreatment recovery. Another requirement for validation trials is the definition of processing parameters in terms of product and processing temperatures, in addition to pressure. The cycle time must also be defined along with compression and decompression rates. The type and integrity of packaging should be provided.

2.6.2 Canada

Within Canada, foods that are resulting from a process not previously used for food and that can cause the food to undergo a major change are considered "Novel Foods." "Major change" generally means a change in the food outside the natural variation including composition, structure, or nutritional quality of the food or its physiological effects; the manner in which the food is metabolized in the body; or microbiological safety, the chemical safety, or the safe use of the food.

Foods and food ingredients treated with HHP have been sold in the Canadian market place since at least 2004 when Health Canada assessed their safety on a case-to-case basis (Health Canada, 2004). Since that time, Health Canada has assessed a total of nine HHP-treated foods and food ingredients of various compositions including RTE meats, raw meats, fruit and vegetable-based juices/smoothies, egg products, and other spreads under a range of treatment pressures (500–600 MPa) and treatment times. HHP-treated foods were reviewed by the Health Canada until 2016 when HHP products were no longer considered a Novel Food if similar to previously approved products (Health Canada, 2016). Yet, new products, or those that are processed under different processing regimes (e.g., reduced pressure or time), do require premarket approval by the Health Canada. In addition, HHP-treated foods sold in Canada must comply with all applicable legislative and regulatory requirements including validation requirements for HHP treatment with a purpose of pathogen reduction where HHP is a critical control point and for reconditioning. As an example, HHP at 600 MPa and maximum 9 min is approved as postlethality and postpackaging intervention in RTE meals for the control of *Listeria* spp.

2.6.3 European Union

In the EU, the regulatory authorities of the 28 member state countries use their own regulatory framework to guarantee that food produced, sold, and consumed within their jurisdiction is safe. In 2004, European Food Safety Authority (EFSA) was established to provide support to EU institutions and the member states on independent scientific advice concerning food safety (European Comission, 2004). The EU Novel Foods Regulation was implemented in 1997 to provide premarket evaluation of new "novel" products. According to New Novel Food Regulation EU 2015/2283, "food produced from a production process not used for food production within the EU before 15th May 1997, which gives rise to significant changes in the composition or structure of a food, affecting its nutritional value, metabolism or level of undesirable substances" is considered Novel Food. The production processes include HHP, UV, and other new technologies. Currently, there are no definitive regulations in the EU related to HHP-treated foods but are considered novel by the EFSA. To date, the EU commission has approved only a narrow range of products such as pasteurized fruit-based preparations at pH 3.2–4.2 to achieve a target shelf life of 60 days at 5°C (41°F) maximum with no defined HHP process parameters. Importantly the commission ruled that HHP was a direct alternative to the thermal pasteurization treatment of 85°C (46°F) for 10 min that was traditionally used for fruit-based products with a shelf life of >10 days. In 2001, EU authorities delegated to National Authorities the legal status of HHP-treated foodstuffs, according to the data provided by the manufacturers. Then later, Spain deliberated that HHP-pasteurized cooked hams were not considered as novel foods; Food Standards Agency, United Kingdom, issued approval to commercialize HHP fruit–based products. and UK Advisory Committee on Novel Foods and Processes considered that "HHP per se is no longer considered a novel process." In 2010, The French National Agency of Health Security of Food, Environment and Labour stated that HHP at 500–600 MPa (72,500–87,000 psi for 3–5 min), at refrigerated or ambient temperature did not lead to significant changes in the nutritional value, metabolism, or level of undesirable substances in foods. In each case, it was successfully argued that although HHP is a novel process, it does not produce novel foods or changed the product composition. Therefore approval was given provided the HHP process was sufficient to satisfy the microbiological criteria and equipment used complies with the Pressure Equipment Directive (97/23/EC).

EU adopted rules regarding Food Hygiene to cover all stages of processing, distribution, and placing on the market of food intended for human consumption. The harmonized food hygiene legislation applies compulsory, self-run, verification programs to industry and combines them with HACCP principles. The legislation is designed to result in good practice guidelines for the EU food industry, together with self-run verification programs. Important features of the new regulation include microbiological standards to reduce foodborne illness for food products, with set criteria for *Salmonella* and *Listeria* levels.

2.6.4 United States

The regulatory requirements within the United States for HHP-treated foods are divided between the US FDA and USDA depending on the product type. Currently, each product is considered on an individual basis to ensure the appropriate risks have been identified and mitigated by HHP. HHP-treated foods are not considered novel and hence do not need to go through premarket approval. However, there is a need to demonstrate that a designated HHP treatment achieves a target log reduction of pathogens of concern. With respect to processed meat, the USDA FSIS published a directive in 2012 identifying HHP as an intervention to control *L. monocytogenes*, thereby fulfilling Alternative 1 or Alternative 2 within RTE meat production. HHP can also be applied to control *E. coli* O157:H7 on beef trim and *Salmonella* on poultry.

Also, in the United States, there is a need to demonstrate the efficacy of the HHP process to control pathogens of concern as part of the entire HACCP plan. The FDA requires processors of juice to develop and implement HACCP systems for their processing operations (21 CFR 6138). The type of product, the process, and the number and nature of potential hazards and hazards that are reasonably likely to occur will determine the extent of a hazard analysis. Both domestic and foreign juice processors that ship juice products to the United States are subject to this regulation.

With the introduction of FSMA, there is a requirement to validate HHP to ensure the specific process conditions support a 5-log reduction of the relevant pathogens. According to the law, validation of HHP used as a kill step is required in order to make sure the HHP treatment is at a level of effectiveness to ensure food safety objectives. The hazard analysis is mandated by FSMA regulations (21 CFR 117) and serves to identify the pathogens of significance for a specific food and its specific process or control measure. Moreover, it will also be required to provide the processing and product parameters along with the validation report. FSIS also provided specific log reductions of pathogens required to be achieved for RTE meats. Specifically, 5-log reduction for *L. monocytogenes*, *Salmonella*, and *E. coli* O157:H7. For RTE poultry products, there is a need to demonstrate a 7-log reduction of *Salmonella*, and for raw beef, a 5-log reduction of *E. coli* O157:H7 (USDA/FSIS, 2005).

The summary of current international regulations for HHP-treated foods is presented in Table 2.7.

2.7 Microbiological challenge testing

When conducting challenge studies, developing new products, or modifying existing formulations, the microbiological loading of the raw material shall be considered. Raw materials must be protected from contamination, handled hygienically, and stored in a manner that will minimize deterioration. During testing, the procedures that will be used do not change raw materials or a formulation (including supplier, type, or addition rate) and represents the actual product as close as possible.

After the potentially hazardous microorganisms have been identified and information about their presence in raw food materials has been obtained, the effects of processing and storage can be tested by microbiological challenge testing (MCT), storage tests, and even by mathematical models for estimation of the number of organisms expected to be present in the product after processing, transportation, storage and consumption, and so on. The relevant pathogen can have a historical association with the product. suggested by an advisory group and/or written within regulations.

An MCT can be defined as a study where product and process performance are evaluated from a microbiological efficiency point of view to reduce target pathogenic or natural spoilage organisms in a worst-case scenario of processing.

Table 2.7 Summary of international regulations for HHP-treated foods and validation.

Country	HHP regulations to comply			
	Novel foods	Available approvals	HACCP	Validation requirement
Australia	Traditional food	HHP-treated yogurt for shelf-life extension	Required for high-risk businesses	Log count reduction achieved is reflective of the expected levels.
New Zealand	Not novel Food	HHP-treated foods that do not require prescribed pasteurization step or used in addition to pasteurization		Shelf-life studies beyond storage time. Temperature is a critical processing parameter.
Canada	Regulated as novel foods under Division 28, Part B of the Food and Drug Regulations	Nine approved processes Not novel process since 2016 for similar to approved foods and process	Not a requirement except for meat and poultry Voluntarily accepted by industry	A requirement for processes where HHP is a critical control point.
European Union	Regulated as novel foods by EFSA	Fruit preparation by EFSA HHP-treated cooked ham, Spain Fruit-based products, United Kingdom All products at 600 MPa (87,000 psi), 3–5 min, France	Compulsory self-run verification programs	Self-run validation
United States	No	No	Requirement for juice, seafood meat, and poultry	FSMA validation is a requirement when HHP is a kill step (can be literature confirmation for similar products).

The purpose of validating HHP can be its application for achieving product

- *safety*, that is, reducing the vegetative pathogenic microorganisms of concern by a known amount and maintaining its safety for the duration of its shelf life.
- *shelf-life extension*, that is, ensuring that the product remains microbiologically and chemically suitable for the duration of its storage.

In terms of evaluating process capability to inactivate pathogenic organism(s), tested product is inoculated with a known number of cells of preselected strains, and the level of inactivation is examined. In contrast, a microbiological shelf-life study simply examines the response as a growth of microorganisms that happened to contaminate the product at the time of conducting the study.

Validation should be undertaken to confirm that HHP parameters such as the holding pressure, time, and temperature are effective at reducing the microorganisms of concern to an acceptable level when applied to the test product.

Validation may be undertaken to demonstrate that the information used to develop the process from the following sources will produce a safe and suitable product:

- HHP parameters that have been previously validated on similar products;
- HHP parameters from reference material, reputable scientific literature, codes of practice, or international standards and that are directly applicable to test product;
- HHP parameters from equipment suppliers and/or HHP experts that are directly applicable to test product;
- HHP parameters that have been developing for test products or applications that have not previously been validated.

When carrying out HHP validation, the following steps can be recommended:

1. Complete the preliminary work to develop your process and product including formulation, composition, process flow, hazard identification, analysis and control point determination, shelf-life requirements, etc. Information about process flow should include all process steps, including the steps prior to the HHP. Critical limits or other limits for the pre-HHP steps should be provided.
2. Carry out the validation of any other aspects of the process (e.g., packaging) that need to occur before the HHP step is validated.

3. Pressure-transmitting fluid: You must warrant that water that is used as pressure-transmitting fluid that comes into direct contact or indirect contact with the product is suitable water.
4. Carry out HHP validation.
5. Process and product parameters that are critical to food safety and shelf life should be identified and managed, including identifying those that would require revalidation if changed.
6. Transfer the validated parameters to your operating procedures.
7. Keep all records generated for assessment by an evaluation and as a record of the work you have completed.
8. The validation needs to be comprehensive and include several repetitions to ensure confidence in the results.
9. Trials carried out to validate an HHP should be repeated if a change is made to the ingredients supply, formulation, pre-HHP steps, packaging, or HP process that could impact food safety.

2.7.1 Shelf-life validation

Shelf life is the time that the product remains safe and suitable for consumption. These tests are conducted to provide justification for setting the product shelf life in relation to microbiological spoilage of the product and often should include other quality issues or deterioration due to chemical factors. The following information about product and process has to be collected during validation trials and included in the report.

2.7.1.1 Product

Microbiological regulatory or operator-defined limits to be met in the product:

- Product formulation
  - o *List ingredients and amounts added*
- Product characteristics
  - o pH; water activity A_w; content of salt/sugar and fat
  - o Other (list)
  - o Weight
- Packaging
  - o Type, size, and form of packaging, materials compliance with regulatory requirement, headspace, sealing
- Intended shelf life and storage temperature
- Spoilage microorganisms of concern and limits to be met
  - o Yeast, mold, and aerobic plate count

2.7.1.2 Process

- Full process details
  - o including preparation, filling, closing, holding, HHP, cooling, storage, etc., containing all processing parameters and limits. This may be provided as a process flow.
  - o HACCP plan or equivalent.
- Proposed HHP test parameters
  - o Initial product temperature before the HHP step.
  - o Hold time at pressure.
  - o Process pressure
  - o Compression and decompression profiles.
  - o Other (e.g., number of cycles, temperature of processing water).
  - o Source of proposed HHP parameters.

2.7.1.3 Test

- Description of HHP system
  - o Details of the HHP equipment used for the trials including loading and unloading and preparation steps.
  - o Is this the same HHP unit that will be used for commercial production?
  - o Facility where HHP is located and equipment details (e.g., make, model, size).
  - o Calibration information.
- Processing conditions.
  - o Product should be processed under worst-case conditions.
- Storage conditions
  - o Optimum, for example, storage as on the label for entire shelf life.
  - o Realistic, for example, repeated short periods of elevated temperatures.
  - o Worst case, for example, poor temperature control in a domestic refrigerator or variable temperature control.
- Sampling for shelf-life studies
  - o Studies should be repeated on at least three batches to identify variability within and between batches. Ideally, these will be on different dates with different batches of ingredients.
  - o Further trials will be needed if the shelf life is not met, or the shelf life must be reduced.
  - o At each testing period, it is recommended to test three to five replicate samples.

  - o Sufficient samples should be stored so that testing can be carried out beyond the stated shelf life, for example, to ensure that any surviving cells that may grow during the shelf life are detected.
  - o A full set of samples will be needed for each storage condition tested. Determine the number needed from the sampling program plus a few extra samples.
  - o If your study requires to testing non-HHP treated controls throughout the shelf life, a sufficient number of controls have to be stored as identified in the sampling program.
- Analyzing data, report, and recommendations.
  - o Validated HHP parameters transferred to operating procedures, including procedures to ensure correct shelf life is applied.

2.7.2 Safety validation

The objective of safety validation is to demonstrate that the HHP will inactivate vegetative pathogenic microorganisms to a required level. Inoculation of food with a known concentration of target pathogenic microorganisms can be used for two purposes: to assess the growth potential of the pathogen in food under specific circumstances and to assess the efficiency of HHP inactivation ability. Depending on the HHP treatment parameters used (e.g., pressure, time, and temperature) combined with the characteristics of the product (e.g., A_w, pH, salt/sugar/fat content), and packaging, the effect on inactivation can vary. Some pathogenic organisms may not be completely killed, but only injured. These organisms may not be detected/enumerated through the analyses performed just after the HHP treatment, especially when selective media are used, but the conditions of storage during shelf-life may allow the recovery of these injured cells and therefore their detection at some stage during the shelf life.

It is important to know if the product and packaging characteristics and storage conditions support the recovery and growth of the target microorganism during the shelf life. If the answer is no, an assessment of log reduction just after the HHP treatment may be sufficient. If the answer is yes, the log reduction needs to be assessed during the product's shelf life. Given this, it is likely that there is a need to demonstrate that the pathogens remain at an acceptable level for the shelf life of the product and therefore product samples will also need to be produced and stored for the shelf life of the product to confirm this.

The information collected about product, process, tests, and storage conditions should be similar to shelf-life studies. However, inoculations studies have to be carefully planned considering organism selection, growth, inoculation, and recovery.

1. Identification of pathogens to be inactivated by the HHP and the reduction to be achieved and/or limits to be met. Confirming that these limits will be met by HHP is the purpose of this validation.
2. Cocktail of microbiological strains should be used for each pathogen and justification for their selection. The strains that should be relevant to the product (those isolated from similar products), include a reference strain, human-related strains, and also pressure-resistant strains. Three to five of the most pressure-resistant strains for the food matrix are recommended.
3. Inoculum preparation should be managed considering the following questions:
   a. Growth stage of cells prior to inoculation that represents realistic → worst case
      i. Methodology should include culture preparation and maintenance.
      ii. Stationary-phase cells preferred.
      iii. "Persisters"/viable but non-culturable *VBNC) can be considered.
   a. Cell adaptation
      i. Determine if the cells can die off in the product.
      ii. Can be adapted to temperature, pH, preservatives, or contamination from ingredients/product.
      iii. Non-adapted: represent environmental contamination.
   b. Inoculation level
      i. Must be high enough (at least 10^6 cfu/mL) to be detectable to see desired response of inactivation.
      ii. Must be low enough that you do not bias outcome.
4. Recovery of microorganisms
   a. Microorganisms are generally/routinely recovered and enumerated from foods using traditional, culture-based methods. The goal is to make sure to use the correct methods for microbe's recovery and enumeration and that the test is repeated in a reasonable number of times to increase confidence.
   b. Recovery media must be selected that supports the growth of target microorganism, and it does not cause stress and optimizes

recovery of sublethally injured cells. The use of selective media is not recommended because the antibiotics/selective agents can inhibit the target organisms if stressed.

c. The recovery conditions should also support the growth of the target organism so the temperature and atmosphere must be precise for the bacteria, and injured cells have sufficient time to grow.

Table 2.8 contains recommendations on the selection of target organisms and inactivation levels for some foods and beverages.

LCR of the target organisms achieved by the process has to be determined and reported as

LCR = $\log_{10}$ cfu/mL of non-HHP **treated control (average) − $\log_{10}$ cfu/mL of** HHP **treated (average**)

Validated HHP parameters are transferred to the operating procedures if the required LCR or other performance criteria have been achieved. Data will also be required showing that the process undertaken does not damage physical and quality characteristics of the product. This is required for spoilage and needed for extending the shelf life.

2.7.3 Surrogate organisms

Given the restrictions on using actual pathogens in validation trials, there is a dependence on identifying suitable surrogates or indicator microbes. An important difference between surrogates and indicators is that the latter occurs naturally in the product and the former has a higher bartolerance compared to the most resistant relevant pathogen.

To obtain quantitative data to support the development and validation of new process, it is necessary to use microbial surrogates which are nonpathogenic species and strains that have demonstrated an equivalent response to HHP treatment than the equivalent pathogenic species and strain. The NACMCF (2010) guidelines also recommend the use of surrogates for validation of microbial efficacy of HHP treatment and establishing safe processing conditions. The choice of surrogate for HHP validation purposes needs to be justified and documented. If no direct relevant published information is available, then studies need to be conducted to establish the validity of using a particular surrogate pathogen—process product parameter combination. It should be considered that a surrogate strain that works well to predict the response for one type of process may not be appropriate for a different type of process (Margosch et al., 2004).

Table 2.8 Target pathogenic organisms for HHP treatment of some food and beverages.

Product	pH	Recommended	Required	Storage
		Pathogen of concern	Log count reduction	
Milk	pH > 4.6	*Campylobacter* spp.	5	Refrigerated
		Listeria monocytogenes	5	
		Shiga toxin–producing strains of *E. coli*	5	
		Salmonella spp.	5	
		Staphylococcus aureus	5	
		Mycobacterium avium subspp. *paratuberculosis* (MAP) (surrogate for *Mycobacterium bovis*)	6	
		Bacillus cereus	No growth	
Raw fruit/vegetable juices; fruit/vegetable nectars; flavored water with fruit/vegetable juices; fruit/vegetable puree, coconut water	pH < 4.6			Refrigerated
	Apple	*E. coli* O157:H7	5	
	Orange	*Salmonella*	5	
	Other fruit and vegetable	*L. monocytogenes* and *Cryptosporidium*	5	
	pH > 4.6	*Clostridium botulinum*	5	
RTE meat and poultry	pH = 4.6	*Listeria*	> 5	Refrigerated
		Salmonella		
		E. coli O157:H7		
Guacamole, hummus, salsa, dip, spreads, marinade, salad dressing	pH < 5.0	*L. monocytogenes*	5	Refrigerated

Sheen et al. (2015) have suggested the use of a nonpathogenic *E. coli* strain, *E. coli* O128 B-59801, as a surrogate for pathogenic *E. coli* strains. This strain had similar pressure resistance to inactivation measured by *D* value at 350 MPa and precompression temperature of 4°C, with *D* value of 39 Shiga toxin—producing strains of *E. coli*.

A heat-resistant beef isolate of *E. coli* AW1.7 has been identified in the literature as one of the most pressure-resistant vegetative bacterial isolates characterized to date (Liu et al., 2015). *E. coli* AW1.7 was found to have comparable resistance to a pressure-resistant mutant *E. coli* isolate (LMM1030) in poultry meat (Liu et al., 2015). Further work would be required to establish if this could be a worst-case surrogate for all vegetative pathogen inactivation using HHP.

Garcia-Hernandez (2015) developed and validated a surrogate strain cocktail to evaluate bactericidal effects of pressure on verotoxigenic *E. coli* (VTEC). It was reported that all highly heat-resistant strains were also pressure resistant but not all pressure-resistant strains were also heat resistant. The resistance to pressure treatment at 600 MPa and 20°C of a cocktail of five nonpathogenic strains of *E. coli* was compared to a 5-strain cocktail of pressure-resistant VTEC. HHP treatments were made in ground beef containing 15% fat. This study identified and validated a cocktail of nonpathogenic *E. coli* to reliably indicate the survival of VTEC after HHP treatment of food.

L. innocua has been often recommended as a surrogate for *L. monocytogenes* although the high intra-strain variability of the genera requires further studies to be performed. A further problem is that *L. innocua* is applied as a hygiene indicator that would limit application in commercial processing facilities

Waite-Cusic et al. (2011) identified a surrogate for *L. monocytogenes* for HHP testing among four *Lactobacillus* spp., a *Pediococcus* sp., and a *L. innocua* strain. These strains were compared with nine *L. monocytogenes* strains, with regard to sensitivity to pressure at 500 MPa for 1 min. *L. monocytogenes* V7 and OSY-8578 were among the most resistant strains to HHP treatments, and thus are candidate target pathogenic strains. *Lactobacillus plantarum* ATCC 8014 demonstrated similar or greater resistance compared with the target organisms; therefore, the bacterium was proposed as a surrogate of *L. monocytogenes* under the conditions tested in the study.

In the case of viruses, MS2 bacteriophages and Tulane virus have been suggested as surrogates for norovirus although studies have illustrated comparison for lower HHP tolerance than the pathogen (Horm et al., 2012; Li et al., 2013).

An alternative to surrogates is to apply predictive models constructed using pathogens inoculated into the relevant food matrix, then treated with HHP. Models have been generated to predict the inactivation kinetics of pathogen in foods although these have been found to be product-specific and likely the strain used to generate the experimental data (Hereu et al., 2012).

2.8 Commercial HHP trials

The following sections provide an overview of commercial HHP challenge trials using various food commodities in relation to the reduction of model pathogens (*Salmonella*, *L. monocytogenes*, and toxigenic *E. coli*). From the data it can be seen that even foods within the same category can differ with respect to the efficacy of HHP that highlights the need to validate individual products rather than category.

2.8.1 Juices

HHP has been applied to a range of juices varying with respect to composition and level of acidity. The specific pH and formulation of juices were not provided presumably due to confidentiality issues. Yet, it was evident that irrespective of the acidity level there was a >5 log cfu reduction of relevant pathogens (*E. coli* O157:H7, *Salmonella*, and *L. monocytogenes*) with pressure treatments of 540–600 MPa and processing times of 30 s or greater. On occasions when <5-log CFU reduction was reported it was noted that there was an initial reduction in pathogen levels attributed to the acidity of the product. There is uncertainty if such log reductions attributed to formulation hurdles should be included in the overall log reduction. For example, a product donated Apple Berry Fiber supported 5-log cfu reduction due to the acidity of the beverage in the absence of HHP treatment. Although the reduction in LCR by pressure was <5 log cfu, it would be problematic to separate out the degree of lethality imparted by the acidity of the product and HHP. To address the issue of having less than 5-log cfu/mL in the controls prior to processing, the inoculation level was increased to up to 8-log cfu/mL. Yet the impact of inoculation level on the inactivation kinetics remains to be elucidated.

From the pooled data it was evident that the barotolerance of the pathogens tested was comparable in low- or high-acid juices (Fig. 2.3). The large variation in log reductions in high acid juice can be attributed to the relative die-off rates due to juice acidity. For low-acid juices there

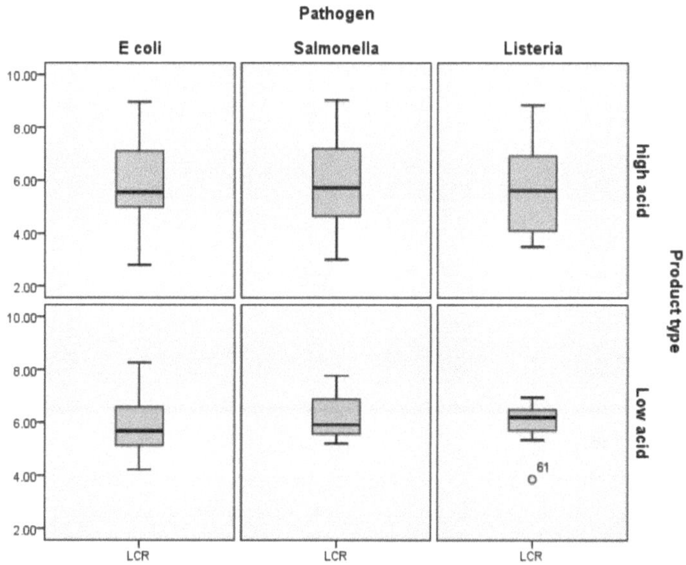

Figure 2.3 Boxplots of pooled data taken from commercial validation trials using various juices inoculated with *Escherichia coli* O157:H7, *Salmonella*, and *Listeria monocytogenes*. The juices (61 types) were subdivided into high acid (pH < 4.5) or low acid (pH > 4.5) with log count reduction obtained at 600 MPa (11°C, 30–180 s) being plotted.

was a relatively low degree of variation in pressure resistance for *Salmonella* and *Listeria* compared to *E. coli* O157:H7. Yet, overall there was no significant difference between the log reductions observed in HHP low- or high-acid juices although residual pathogens were commonly found in the former. This was illustrated with coconut water that was classified as low acidity by the virtue of having a pH 5.8. Although a 5-log reduction could be achieved, low levels of survivors were frequently encountered with complete elimination of target pathogens being rarely encountered. The addition of carrot juice into beverage blends also appeared to have a protective effect with lower log reductions being obtained compared to high-acid fruit juices tested. It is unclear if a constituent within carrot juice exhibited a protective effect or was a case of diluting the acidity of the product.

A further finding from commercial trials is the negligible impact of processing time (30–300 s) on the pathogen log reduction. However, it should be noted that treatment times in the order of 180–300 s were required to inactivate low numbers of residual survivors. The level of

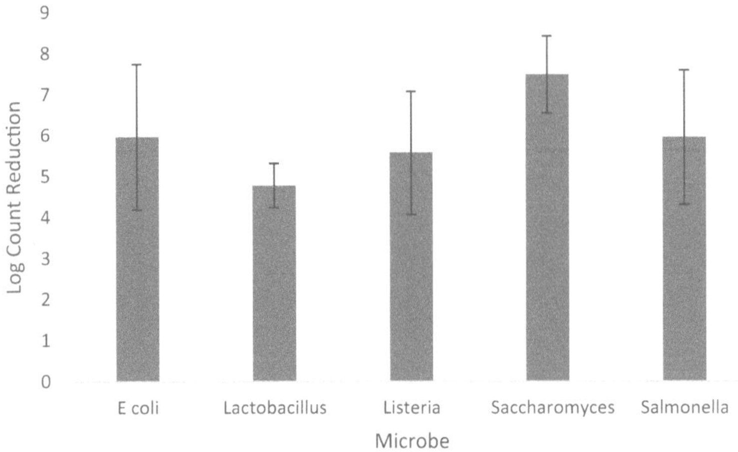

Figure 2.4 Log count reduction of different microbes in high-acid juices. Data were extracted from the log reductions obtained by treating high-acid juices with 600 MPa for 120–180 s. The initial inoculation level in all cases was 7-log cfu/mL.

applied pressure did have an impact on the log reductions being observed although the test range was typically between 540 and 600 MPa. As a general rule, 600 MPa was required to support a 5-log reduction with lower pressures falling short of the target.

In addition to pathogenic challenge studies, the efficacy of HHP to inactivate potential spoilage microbes was also undertaken. Specifically, the inactivation of *Lactobacillus* and *Saccharomyces* in high-acid juices was reported. From the pooled data, it was evident that *Saccharomyces* was most sensitive to HHP treatment with *Lactobacillus* exhibiting higher tolerance (Fig. 2.4). With regards to the latter, it was noted that *Lactobacillus* exhibited higher barotolerance than the pathogens tested suggesting the bacterium could be a candidate for verification trials within commercial processing facilities.

In conclusion, the commercial data from the HHP of juices provided evidence that high-acid type beverages should be considered as low risk and compatible with HHP. However, low-acid juices may represent a hazard due to the lower antimicrobial effect of pressure against microbes and potential for outgrowth to occur if temperature abuse occurs during storage.

2.8.2 Meats

The majority of data generated from commercial validation studies is directed towards the HHP of deli meats. Still, the efficacy of HHP to inactivate pathogens in raw and other processed meats has also been tested.

The data from the commercial trials were subdivided into different pathogen types with the meat products being subdivided into cured and noncured products. The formulation of the different meat products were not made available and it was possible that the presence of antimicrobial ingredients (for example, diacetate) could have influenced the efficacy of the high pressure process.

From the pooled data, it was evident that the log reduction of *E. coli* O157:H7 on cured meat products was lower than that on noncured product (Fig. 2.4). This was less evident with *Salmonella* and *L. monocytogenes*. A further noteworthy point was the relatively large variation in pressure tolerance depending on product type (except for *E. coli* O157: H7 on cured meats) (Fig. 2.5). With respect to noncured meats, the lowest log reduction recorded was for a beef deli product with a treatment of 600 MPa for 120 s supporting a 2.25 LCR. The same product also resulted in the lowest log reduction for *Listeria*. However, for *Salmonella*, the lowest log reduction (3.36-log cfu) was obtained for a turkey roll product. The results would suggest that both formulation hurdles and pathogen type influence the efficacy of HHP. In raw meat (ground beef) that had a higher water activity ($A_w = 0.99$), the observed log reductions

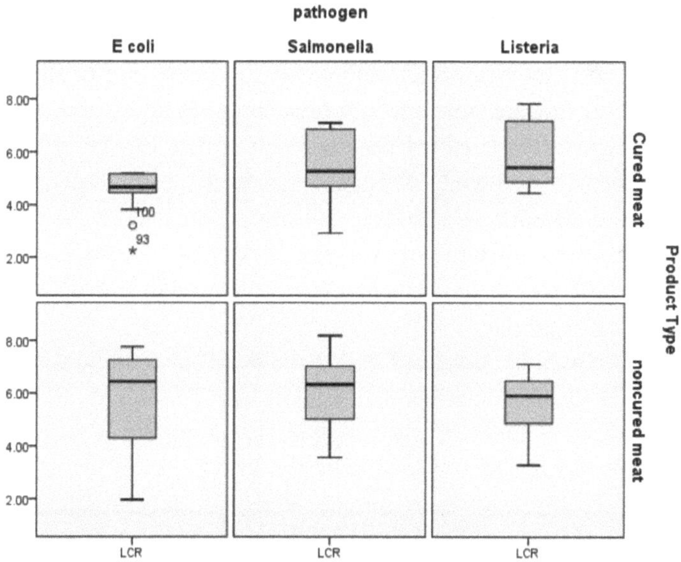

Figure 2.5 Boxplots of pooled log count reduction data taken from HHP trials of inoculated cured and noncured meats. The pressure treatment applied was 600 MPa for 120–350 s at 11°C.

achieved using HHP were lower compared to deli meats with corresponding A_w and pH. *E. coli* O157:H7 appeared to exhibit enhanced barotolerance on ground beef compared to either *Salmonella* or *Listeria*.

A common feature of HHP of meat products was the recovery of residual pathogens that could not be eliminated by extended processing times (up to 9 min) or if samples were screened 72 h postprocessing. Consequently, there is a potential for potential pathogen outgrowth to occur if the product is temperature abused and additional hurdles are not in place such as sodium diacetate.

In summary, the efficacy of HHP of meat should be considered a pathogen reduction step although complete elimination is rarely observed. HHP is less effective as a pathogen reduction step on raw meat or on those products formulated to have low water activity.

2.8.3 Deli salads, hash browns, herbs, and gravy base

Deli salads treated have been successfully treated with HHP with total elimination of pathogens being observed in certain cases. HHP is effective with low pH salads where the dressing most likely supports the highest log reduction. Yet, it was noted the efficacy of HHP to inactivate pathogens was dependent on the salad type and pathogen tested. For example, *Salmonella* exhibited barotolerance when introduced into apple pecan salad but sensitive on pasta salad. Such differences are likely due to the ingredient components or other intrinsic factors such as pH and water activity.

Hash browns are a breaded potato product that can be treated with HHP to support >5 log cfu reduction of model pathogens. However, hash browns are unique in that from all the products tested it appears to assist the recovery/growth of pathogens during post-HHP storage. Indeed, *L. monocytogenes* underwent repair and growth during post-HHP storage at 4°C for 20 days.

Hard-boiled whole and diced eggs have been successfully treated with HHP in terms of pathogen reduction. The log reductions of pathogens achieved was independent of whole egg or diced egg being processed.

Gravy base has also been tested although low LCRs of pathogens were observed due to the low water activity of the product. A single trial has been performed with leafy herbs (basil) with successful reduction of endogenous coliforms to below the limit of detection (<1 log cfu/g). The results would suggest the potential to treat a diverse range of foods and ingredients using HHP.

2.9 Responsibilities for contract of HHP facility

In order to conduct HHP validation study, the contract processor must be correctly defined, agreed, and controlled in order to prevent noncompliance that could result in food safety and/or shelf-life issues.

First, there must be a written agreement between the manufacturer and the HHP processor which clearly establishes the duties of each party. Technical aspects of the agreement should be drawn up by competent persons suitably knowledgeable in food manufacturing and GMP. All cycle parameters for processing must be included. The agreement should specify the way in which the authorized person releasing the batch for sale ensures that each batch has been processed and checked for compliance with the requirements for food safety.

The food manufacturer is responsible for assessing the competence of the HHP processor to carry out the work required. The food manufacturer should ensure that the processor is fully aware of any problems associated with the product or packaging which might pose a food safety hazard. The manufacturer should ensure that all processed products delivered to the processor have been released by an authorized person. A manufacturing record should record transport to the HHP processor, storage conditions at the processor, processing times and cycle numbers, and the return transportation.

The processor must have adequate premises and equipment, knowledge and experience in HHP, and competent personnel to carry out the validation work ordered by the manufacturer.

Records of the relevant HHP treatment cycles should be available to the food manufacturer; have a clear link between the manufacturers' batch number and the processors' cycle number(s); and ensure that any records relevant to assessing the quality of a product in the event of complaints or a suspected defect are accessible and specified in the complaints handling or recall procedures of the manufacturer.

The examples of HHP process records may include:

- processing date and time;
- product name, code, and/or other identification;
- formulation;
- pack size;
- product parameters (pH, A_w, product composition, etc.);
- approximate number of containers in the load;
- initial product temperature;

- initial temperature of pressure-transmitting fluid (water);
- compression time;
- target pressure;
- HT;
- decompression time;
- results of seal checks;
- results of final product testing; and
- water suitability checks.

A report of the validation work must be documented by the suitably skilled person.

The validation report must be kept:

1. while the product is in production.
2. for additional years or the shelf life of the last product produced, whichever is longer, if its production ceases.
3. If the validation report is replaced, the obsolete report must be kept.

2.9.1 Deviation from the validated process

The immediate action shall be taken if there is a process deviation that could impact on food safety or suitability, including if any regulatory or operator-defined limit is not met. Any product lot(s) affected by a deviation that could impact on food safety should be identified, segregated, and detained pending the outcome of an assessment by a suitably skilled person. Then, a record of the assessment and corrective actions taken must be prepared. The record should be appropriate to the nature of the deviation and include:

1. date and time of deviation;
2. description of the deviation, including the processing records;
3. description of affected product, including batch identifiers and quantity;
4. corrective action taken, including restoration of control, product disposition, and prevention of recurrence;
5. records of any tests undertaken.

2.10 Future prospects and gaps in HHP technology development

Even though the first reports of HHP for preserving foods were published over 100 years ago, it has only been recently applied in mainstream food processing. Nowadays the use of HHP is growing and more than 500 units have been installed worldwide. Though the investment cost in HHP technology has not changed substantially, improvements in the machines' configurations and components have translated to higher productivity and lower production costs.

The transition from concept to commercial processing of HHP was very rapid with the consequence of leaving many knowledge gaps including process validation. This is especially relevant in terms of predicting pathogen inactivation along with how both the intrinsic and extrinsic factors affect the process efficacy. In relation to this, one of the most striking features is the dependency on the product whereby HHP can reduce spoilage enzymes in certain matrices but increases in others. This product dependency is also observed in microbial reduction that makes the inactivation process challenging to predict along with model and complicates process validation. Based on this, HHP validation approaches can differ for the groups of products with the special attention on product behavior under pressure, choice of organism of concern and its surrogate, packaging material and design, and thorough characterization of equipment performance. Also, there are knowledge gaps that can be addressed in terms of HHP scaling up and other engineering aspects that include process and product uniformity, temperature control within the product and within the vessel. In general, the high amount of published information should be used in systematic analysis and used for predicting microbial reduction in different food matrices and samples with a variety of shapes, geometry, and packaging configuration and should be used in the optimization of HHP treatment conditions.

References

Abe, F., 2007. Exploration of the effects of high hydrostatic pressure on microbial growth, physiology and survival: perspectives from piezophysiology. Biosci. Biotechnol. Biochem. 71, 2347–2357.

Aertsen, A., Vanoirbeek, K., De Spiegeleer, P., Sermon, J., Hauben, K., Farewell, A., et al., 2004. Heat shock protein-mediated resistance to high hydrostatic pressure in Escherichia coli. Appl. Environ. Microbiol. 70, 2660–2666.

Alba, M.d., Bravo, D., Medina, M., 2013. Inactivation of Escherichia coli O157:H7 in dry-cured ham by high-pressure treatments combined with biopreservatives. Food Control 31 (2), 508–513.

Ananta, E., Knorr, D., 2004. Evidence on the role of protein biosynthesis in the induction of heat tolerance of Lactobacillus rhamnosus GG by pressure pre-treatment. Int. J. Food Microbiol. 96, 307–313.

Anese, M., Nicoli, M.C., Dallaglio, G., Lerici, C.R., 1995. Effect of high-pressure treatments on peroxidase and polyphenoloxidase activities. J. Food Biochem. 18, 285–293.

Avsaroglu, M.D., Bozoglu, F., Alpas, H., Largeteau, A., Demazeau, G., 2015. Use of pulsed-high hydrostatic pressure treatment to decrease patulin in apple juice. High Pressure Res. 35, 214–222.
Balasubramaniam, V.M., Ting, E.Y., Stewart, C.M., Robbins, J.A., 2004. Recommended laboratory practices for conducting high-pressure microbial inactivation experiments. Innov. Food Sci. Emerg. Technol. 5 (3), 299–306.
Baron, A., Denes, J.M., Durier, C., 2006. High-pressure treatment of cloudy apple juice. Lwt-Food Sci. Technol. 39, 1005–1013.
Bartlett, D.H., 2002. Pressure effects on in vivo microbial processes. Biochim. Biophys. Acta 1595 (1–2), 367–381.
Bevilacqua, A., D'Amato, D., Sinigaglia, M., Corbo, M.R., 2013. Combination of homogenization, citrus extract and vanillic acid for the inhibition of some spoiling and pathogenic bacteria representative of dairy microflora. Food Bioproc. Technol. 6, 2048–2058.
Bridgman, P.W., 1931. The physics of high pressure. London: G. Bells and Sons.
Bruna, D., Voldrich, M., Marek, M., Kamarad, J., 1999. Effect of processing and storage conditions on patulin content in apple juice concentrate. Czech J. Food Sci. 17, 127–132.
Cano, M.P., Hernandez, A., DeAncos, B., 1997. High pressure and temperature effects on enzyme inactivation in strawberry and orange products. J. Food Sci. 62, 85–88.
Chawla, R., Patil, G.R., Singh, A.K., 2011. High hydrostatic pressure technology in dairy processing: a review. J. Food Sci. Technol. Mys. 48 (3), 260–268.
Considine, T., Patel, H.A., Anema, S.G., Singh, H., 2007. Interactions of milk proteins during heat and high hydrostatic pressure treatments—a review. Innov. Food Sci. Emerg. Technol. 8 (1), 1–23.
Demoly, P., Tanno, L.K., Akdis, C.A., Lau, S., Calderon, M.A., Santos, A.F., et al., 2014. Global classification and coding of hypersensitivity diseases—An EAACI - WAO survey, strategic paper and review. Allergy 69, 559–570.
Denys, S., Hendrickx, M.E., 1999. Measurement of the thermal conductivity of foods at high pressure. J. Food Sci. 64, 709–713.
European Commission, 2004. Evaluation report on the novel food regulation 258/97.
Farakos, S.M.S., Zwietering, M.H., 2011. Data analysis of the inactivation of foodborne microorganisms under high hydrostatic pressure to establish global kinetic parameters and influencing factors. J. Food Protect. 74 (12), 2097–2106.
Food Standards Australia New Zealand, 2004. Review of novel food standard. Initial Assessment Report. 4 p.
Garcia-Hernandez, R., 2015. Variation in resistance to high hydrostatic pressure of *Escherichia coli* (PhD thesis). University of Alberta, Edmonton.
Gayan, E., Torres, J.A., Paredes-Sabja, D., 2012. Hurdle Approach to increase the microbial inactivation by high pressure processing: effect of essential oils. Food Eng. Rev. 4, 141–148.
Goodner, J.K., Braddock, R.J., Parish, M.E., 1998. Inactivation of pectinesterase in orange and grapefruit juices by high pressure. J. Agric. Food Chem. 46, 1997–2000.
Haiqiang, C., 2007. Use of linear, Weibull, and log-logistic functions to model pressure inactivation of seven foodborne pathogens in milk. Food Microbiol. 24 (3), 197–204.
Hao, H.Y., Zhou, T., Koutchma, T., Wu, F., Warriner, K., 2016. High hydrostatic pressure assisted degradation of patulin in fruit and vegetable juice blends. Food Control 62, 237–242.
Hauben, K.J.A., Wuytack, E.Y., Soontjens, C.C.F., Michiels, C.W., 1996. High-pressure transient sensitization of Escherichia coli to lysozyme and nisin by disruption of outer-membrane permeability. J. Food Protect. 59, 350–355.

Hayman, M., Anantheswaran, R., Knabel, S., 2007. Dehydration protects Listeria monocytogenes from inactivation by high pressure processing. Abstracts of the General Meeting of the American Society for Microbiology, vol. 107, p. 517.
Heinz, V., Knorr, D., 2002. Reduction of pathogen microorganisms by high pressure and pulsed electric fields: theoretical and practical considerations. Ernaehrung 26 (3), 102–106.
Health Canada, 2010. Novel Foods [Online] (August 2016).
Health Canada, 2014. Health Canada position - high pressure processing (HPP) is no longer a novel process. Available at https://www.canada.ca/en/health-canada/services/food-nutrition/legislation-guidelines/guidance-documents/position-high-pressure-processing-no-longer-novel-process-treated-food-products-treated-food-products-2013.html.
Health Canada, 2016. Health Canada position – high pressure processing (hpp) is no longer a novel process [Online] (January 2017).
Hereu, A., Dalgaard, P., Garriga, M., Aymerich, T., Bover-Cid, S., 2012. Modeling the high pressure inactivation kinetics of Listeria monocytogenes on RTE cooked meat products. Innov. Food Sci. Emerg. Technol. 16, 305–315.
Hite, B.H., Giddings, N.J., Weakly, C.E., 1914. The effects of pressure on certain microorganisms encountered in the preservation of fruits and vegetables. WVU Agric. Exp. Sta. Bull. 146, 1–67.
Hoover, D.G., Metrick, C., Papineau, A.M., Farkas, D.F., Knorr, D., 1989. Biological effects of high hydrostatic-pressure on food microorganisms. Food Technol. 43 (3), 99–107.
Horm, K.M., Davidson, P.M., Harte, F.M., D'Souza, D.H., 2012. Survival and inactivation of human norovirus surrogates in blueberry juice by high-pressure homogenization. Foodborne Pathog. Dis. 9 (11), 974–979.
Hormann, S., Scheyhing, C., Behr, J., Pavlovic, M., Ehrmann, M., Vogel, R.F., 2006. Comparative proteome approach to characterize the high-pressure stress response of Lactobacillus sanfranciscensis DSM 20451. Proteomics 6, 1878–1885.
Huang, H.-W., Hsu, C.-P., Yang, B.B., Wang, C.-Y., 2014. Potential utility of high-pressure processing to address the risk of food allergen concerns. Compr. Rev. Food Sci. Food Saf. 13, 78–90.
Johnson, P.E., Van der Plancken, I., Balasa, A., Husband, F.A., Grauwet, T., Hendrickx, M., et al., 2010. High pressure, thermal and pulsed electric-field-induced structural changes in selected food allergens. Mol. Nutr. Food Res. 54, 1701–1710.
Juck, G., Neetoo, H., Beswick, E., Chen, H., 2012. Influence of prior growth conditions, pressure treatment parameters, and recovery conditions on the inactivation and recovery of Listeria monocytogenes, Escherichia coli, and Salmonella Typhimurium in turkey meat. Int. J. Food Microbiol. 153, 203–211.
Karatzas, K.A.G., Bennik, M.H., 2002. Characterization of a Listeria monocytogenes Scott A isolate with high tolerance towards high hydrostatic pressure. Appl. Environ. Microbiol. 68, 3183–3189.
Karatzas, K.A.G., Wouters, J.A., Gahan, C.G.M., Hill, C., Abee, T., Bennik, M.H.J., 2003. A spontaneous deletion in the glycine repeat region of the ctsR regulator of Listeria monocytogenes affects piezotolerance, stress resistance and virulence. FEMS Congress of European Microbiologists Abstract Book, 64.
Kim, Y.S., Park, S.J., Cho, Y.H., Park, J., 2001. Effects of combined treatment of high hydrostatic pressure and mild heat on the quality of carrot juice. J. Food Sci. 66, 1355–1360.
Kingsley, D.H., Hollinian, D.R., Calci, K.R., Chen, H., Flick, G.J., 2007. Inactivation of a norovirus by high-pressure processing. Appl. Environ. Microbiol. 73 (2), 581–585.
Kleber, N., Maier, S., Hinrichs, J., 2007. Antigenic response of bovine beta-lactoglobulin influenced by ultra-high pressure treatment and temperature. Innov. Food Sci. Emerg. Technol. 8, 39–45.

Knoerzer, K., Buckow, R., Versteeg, C., 2010. Adiabatic compression heating coefficients for high-pressure processing - a study of some insulating polymer materials. J. Food Eng. 98 (1), 110–119.
Koseki, S., Yamamoto, K., 2006. Recovery of Escherichia coli ATCC 25922 in phosphate buffered saline after treatment with high hydrostatic pressure. Int. J. Food Microbiol. 110, 108–111.
Lado, B.H., Yousef, A.E., 2002. Alternative food-preservation technologies: efficacy and mechanisms. Microb. Infect. 4 (4), 433–440.
Lakshmi, F.U., Chauhan, O.P., Raju, P.S., 2013. Novel non-thermal methods of food processing. Indian Food Ind. 32 (3), 42–49.
Lambert, Y., Demazeau, G., Largeteau, A., Bouvier, J.M., Laborde-Croubit, S., Cabannes, M., 2000. Packaging for high-pressure treatments in the food industry. Packag. Technol. Sci. 13 (2), 63–71.
Lee, J., Kaletunc, G., 2010. Inactivation of *Salmonella Enteritidis* strains by combination of high hydrostatic pressure and nisin. Int. J. Food Microbiol. 140, 49–56.
Li, X.H., Ye, M., Neetoo, H., Golovan, S., Chen, H.Q., 2013. Pressure inactivation of Tulane virus, a candidate surrogate for human norovirus and its potential application in food industry. [Article]. Int. J. Food Microbiol. 162 (1), 37–42.
Liu, Y., Gill, A., McMullen, L.M., Gänzle, M.G., 2015. Variation in heat and pressure resistance of verotoxigenic and non-toxigenic Escherichia coli. J. Food Prot. 78, 111–120.
Lori, S., Buckow, R., Knorr, D., Heinz, V., Lehmacheri, A., 2007. Predictive model for inactivation of Campylobacter spp. by heat and high hydrostatic pressure. J. Food Protect. 70 (9), 2023–2029.
Ma, L., Su, Y.-C., 2011. Validation of high pressure processing for inactivating Vibrio parahaemolyticus in Pacific oysters (Crassostrea gigas). Int. J. Food Microbiol. 144 (3), 469–474.
Margosch, D., Ehrmann, M.A., Ganzle, M.G., Vogel, R.F., 2004. Comparison of pressure and heat resistance of *Clostridium botulinum* and other endospores in mashed carrots. J. Food Protect. 67, 2530–2537.
Metrick, C., Hoover, D.G., Farkas, D.F., 1989. Effects of high hydrostatic-pressure on heat-resistant and heat-sensitive strains of salmonella. J. Food Sci. 54 (6), 1547-&.
Min, S., Sastry, S.K., Balasubramaniam, V.M., 2010. Compressibility and density of select liquid and solid foods under pressures up to 700 MPa. J. Food Eng. 96 (4), 568–574.
Moake, M.M., Padilla-Zakour, O.I., Worobo, R.W., 2005. Comprehensive review of patulin control methods in foods. Compr. Rev. Food Sci. Food Saf. 4, 8–21.
Mohamed, H.M.H., Diono, B.H.S., Yousef, A.E., 2012. Structural changes in Listeria monocytogenes treated with gamma radiation, pulsed electric field and ultra-high pressure. J. Food Saf. 32, 66–73.
NACMCF, 2010. Parameters for determining inoculated pack/challenge study protocols. J. Food Protect. 73, 140–202.
New South Wales Food Authority, 2016. High pressure processing - product considerations in HPP. Available from http://www.foodauthority.nsw.gov.au/_Documents/industry/high_pressure_processing.pdf.
Noma, S., Tomita, C., Shimoda, M., Hayakawa, I., 2004. Response of Escherichia coli O157: H7 in apple and orange juices by hydrostatic pressure treatment with rapid decompression. Food Microbiol. 21 (4), 469–473.
Norton, T., Sun, D.-W., 2008. Recent advances in the use of high pressure as an effective processing technique in the food industry. Food Bioproc. Technol. 1 (1), 2–34.
Palhano, F.L., Gomes, H.L., Orland, M.T.D., Kurtenbach, E., Fernandes, P.M.B., 2004. Pressure response in the yeast Saccharomyces cerevisiae: from cellular to molecular approaches. Cell. Mol. Biol. 50, 447–457.

Palou, E., Lopez-Malo, A., Barbosa-Canovas, G.V., Welti-Chanes, J., Swanson, B.G., 1997. High hydrostatic pressure as a hurdle for Zygosaccharomyces bailii inactivation. J. Food Sci. 62 (4), 855–857.
Patazca, E., Koutchma, T., Balasubramaniam, V.M., 2007. Quasi-adiabatic temperature increase during high pressure processing of selected foods. J. Food Eng. 80 (1), 199–205.
Patterson, M.F., 2005. Microbiology of pressure-treated foods. J. Appl. Microbiol. 98 (6), 1400–1409.
Penas, E., Gomez, R., Frias, J., Luisa Baeza, M., Vidal-Valverde, C., 2011. High hydrostatic pressure effects on immunoreactivity and nutritional quality of soybean products. Food Chem. 125, 423–429.
Pilavtepe-Celik, M., 2013. High hydrostatic pressure (HHP) inactivation of foodborne pathogens in low-acid juices. Int. J. Food Sci. Technol. 48 (4), 673–677.
Prestamo, G., Arabas, J., Fonberg-Broczek, M., Arroyo, G., 2001. Reaction of B-cereus bacteria and peroxidase enzymes under pressures > 400 MPa. J. Agric. Food Chem. 49, 2830–2834.
Ramaswamy, R., Balasubramaniam, V.M., 2007. Effect of polarity and molecular structure of selected liquids on their heat of compression during high pressure processing. High Pressure Res. 27, 299–307.
Ramaswamy, H.S., Riahi, E., Idziak, E., 2003. High-pressure destruction kinetics of E-coli (29055) in apple juice. J. Food Sci. 68 (5), 1750–1756.
Ramaswamy, R., Balasubramaniam, V.M., Sastry, S.K., 2005. Effect of high pressure and irradiation treatments on hydration characteristics of navy beans. Int. J. Food Eng. 1.
Rasanayagam, V., Balasubramaniam, V.M., Ting, E., Sizer, C.E., Bush, C., Anderson, C., 2003. Compression heating of selected fatty food materials during high-pressure processing. J. Food Sci. 68, 254–259.
Rastogi, N.K., Raghavarao, K.S.M.S., Balasubramaniam, V.M., Niranjan, K., Knorr, D., 2007. Opportunities and challenges in high pressure processing of foods. Crit. Rev. Food Sci. Nutr. 47, 69–112.
Robey, M., Benito, A., Hutson, R.H., Pascual, C., Park, S.F., Mackey, B.M., 2001. Variation in resistance to high hydrostatic pressure and rpoS heterogeneity in natural isolates of Escherichia coli O157: H7. Appl. Environ. Microbiol. 67 (10), 4901–4907.
Samaranayake, C.P., Sastry, S.K., 2013. In-situ pH measurement of selected liquid foods under high pressure. Innov. Food Sci. Emerg. Technol. 17, 22–26.
Setikaite, I., Koutchma, T., Patazca, E., Parisi, B., 2009. Effects of water activity in model systems on high-pressure inactivation of Escherichia coli. Food Bioproc. Technol. 2, 213–221.
Sevenich, R., Bark, F., Crews, C., Anderson, W., Pye, C., Riddellova, K., et al., 2013. Effect of high pressure thermal sterilization on the formation of food processing contaminants. Innov. Food Sci. Emerg. Technol. 20, 42–50.
Sheen, S., Cassidy, J., Scullen, B., Sommers, C., 2015. Inactivation of a diverse set of shiga toxin-producing Escherichia coli in ground beef by high pressure processing. Food Microbiol. 52, 84–87.
Shook, C.M., Shellhammer, T.H., Schwartz, S.J., 2001. Polygalacturonase, pectinesterase, and lipoxygenase activities in high-pressure-processed diced tomatoes. J. Agric. Food Chem. 49, 664–668.
Shriver, S.K., Yang, W.W., 2011. Thermal and nonthermal methods for food allergen control. Food Eng. Rev. 3, 26–43.
Sicherer, S.H., Sampson, H.A., 2010. Food allergy. J. Allergy Clin. Immunol. 125, S116–S125.
Simpson, R.K., Gilmour, A., 1997. The effect of high hydrostatic pressure on Listeria monocytogenes in phosphate-buffered saline and model food systems. J. Appl. Microbiol. 83, 181–188.

Smelt, J., Hellemons, J.C., Wouters, P.C., van Gerwen, S.J.C., 2002. Physiological and mathematical aspects in setting criteria for decontamination of foods by physical means. Int. J. Food Microbiol. 78 (1–2), 57–77.
Smiddy, M., Sleator, R.D., Patterson, M.F., Hill, C., Kelly, A.L., 2004. Role for compatible solutes glycine betaine and L-carnitine in listerial barotolerance. Appl. Environ. Microbiol. 70 (12), 7555–7557.
Sonoike, K., Setoyama, T., Kuma, Y., Kobayashi, S., 1992. Effect of pressure and temperature on the death rates of lactobacillus-casei and escherichia-coli. High Pressure Biotechnol. 224, 297–301.
Syed, Q.A., Buffa, M., Guamis, B., Saldo, J., 2014. Effect of Compression and Decompression Rates of High Hydrostatic Pressure on Inactivation of Staphylococcus aureus in Different Matrices. Food Bioproc. Technol. 7 (4), 1202–1207.
Tholozan, J.L., Ritz, M., Jugiau, F., Federighi, M., Tissier, J.P., 2000. Physiological effects of high hydrostatic pressure treatments on *Listeria monocytogenes* and *Salmonella typhimurium*. J. Appl. Microbiol. 88, 202–212.
Toledo, J., Pulido, R.P., Abriouel, H., Grande, M.J., Galvez, A., 2012. Inactivation of Salmonella enterica cells in Spanish potato omelette by high hydrostatic pressure treatments. Innov. Food Sci. Emerg. Technol. 14, 25–30.
Tomasula, P.M., Renye, J.A., Van Hekken, D.L., Tunick, M.H., Kwoczak, R., Toht, M., et al., 2014. Effect of high-pressure processing on reduction of Listeria monocytogenes in packaged Queso Fresco. J. Dairy Sci. 97 (3), 1281–1295.
Toro-Funes, N., Bosch-Fuste, J., Veciana-Nogues, M.T., Vidal-Carou, M.C., 2014. Effect of ultra high pressure homogenization treatment on the bioactive compounds of soya milk. Food Chem. 152, 597–602.
Ueno, S., Shigematsu, T., Hasegawa, T., Higashi, J., Anzai, M., Hayashi, M., et al., 2011. Kinetic analysis of E. coli inactivation by high hydrostatic pressure with salts. J. Food Sci. 76 (1), M47–M53.
USDA/FSIS, 2005. Meat and Poultry Hazards and Controls Guide. Food Safety and Inspection Service, Washington, D.C.
Van Buggenhout, S., Messagie, I., Van der Plancken, I., Hendrickx, M., 2006. Influence of high-pressure-low-temperature treatments on fruit and vegetable quality related enzymes. Eur. Food Res. Technol. 223, 475–485.
Vanlint, D., Pype, B.J.Y., Rutten, N., Vanoirbeek, K.G.A., Michiels, C.W., Aertsen, A., 2013. Loss of cAMP/CRP regulation confers extreme high hydrostatic pressure resistance in *Escherichia coli* O157:H7. Int. J. Food Microbiol. 166 (1), 65–71.
Vogel, R.F., Herdegen, V., Klostermaier, P., 1999. Description of Listeria behaviour under various pressurization strategies. Adv. High Pressure Biosci. Biotechnol. 97–102.
Waite-Cusic, J, Diono, V.H.S., Yousef, A.E., 2011. Screening for Listeria monocytogenes surrogate strains appliable to food processing by ultrahigh pressure and pulsed electric field. J. Food Prot. 74 (10), 1655–1661.
Weemaes, C., Ludikhuyze, L., Van den Broeck, I., Hendrickx, M., 1998. High pressure inactivation of polyphenoloxidases. J. Food Sci. 63, 873–877.
Wemekamp-Kamphuis, H.H., Wouters, J.A., de Leeuw, P., Hain, T., Chakraborty, T., Abee, T., 2004. Identification of sigma factor sigma(B)-controlled genes and their impact on acid stress, high hydrostatic pressure, and freeze survival in Listeria monocytogenes EGD-e. Appl. Environ. Microbiol. 70 (6), 3457–3466.
Woldemariam, H.W., Emire, S.A., 2019. High pressure processing of foods for microbial and mycotoxins control: current trends and future prospects. Cogent Food Agric. 5, 21.
Xu, H., Lee, H.-Y., Ahn, J., 2009. High pressure inactivation kinetics of Salmonella enterica and Listeria monocytogenes in milk, orange juice, and tomato juice. Food Sci. Biotechnol. 18 (4), 861–866.

Zhou, S.Y., Sheen, S.S., Zhao, G.H., Chuang, S.Y., Liu, L.S., 2020. Prediction of Salmonella inactivation in sliced tomato subject to high pressure processing and trans-cinnamaldehyde treatment using selective and non-selective growth media for survival evaluations. Food Control 118, 10.
Zhu, S., Ramaswamy, H.S., Marcotte, M., Chen, C., Shao, Y., Le Bail, A., 2007. Evaluation of thermal properties of food materials at high pressures using a dual-needle line-heat-source method. J. Food Sci. 72, E49–E56.

Further reading

CFIA, 2017. Guidance on food products treated with high pressure processing (HPP) [Online].
Food Standards Australia New Zealand, 2017 Regulation of novel foods, http://www.foodstandards.gov.au/industry/novel/Pages/default.aspx (accessed 02.03.18).
United States FDA Federal Register, 2013. Current good manufacturing practice and hazard analysis and risk-based preventive controls for human food. Federal Register 78 (11), 3646. Available at https://www.govinfo.gov/app/details/FR-2013-01-16/2013-00125 Accessed 11.07.2021

CHAPTER 3

Case study of validation of high hydrostatic pressure processing of fruit and vegetable smoothies in PET bottles

Tatiana Koutchma[1] and Keith Warriner[2,3]
[1]Novel Food Processing Technologies, Agriculture and Agri-Food Canada, Guelph, ON, Canada
[2]Food Science Department, University of Guelph, Guelph, ON, Canada
[3]Center for Public Health and Zoonoses Canada

The objectives of microbial challenge studies are

1. to determine physical and chemical characteristics of smoothies (pH and Brix);
2. to verify that high pressure processing (HPP) treatment of 600 MPa for 3 min supports a 5 log cfu reduction of relevant pathogenic bacteria, such as *Escherichia coli* O157:H7, *Salmonella*, and *Listeria monocytogenes* in inoculated smoothies; and
3. to determine the efficacy of different time—pressure combinations of HPP treatments and the effect of juice formulations on the reduction of model pathogens introduced into juices.

3.1 Product description

Three types of fruit and vegetable-based smoothies will be tested

- Cheeky carrot and smoothie;
- Groovy beet and strawberry smoothie; and
- Cool cucumber and pineapple smoothie.

3.2 Packaging

Two types of round PET bottles will be used: 32 and 4 oz.

Validation of Food Preservation Processes based on Novel Technologies
DOI: https://doi.org/10.1016/B978-0-12-815888-3.00001-4

3.3 HPP unit

- 55-L Hyperbaric HPP unit;
- Maximum operating pressure—600 MPa; compression medium is water.

3.4 Methodology

The bacteria to be tested are five-strain cocktails of *L. monocytogenes*, *E. coli*, and *Salmonella*. The juices to be tested are inoculated with the test pathogen, then transferred to 4- or 32-oz bottles, closed, and placed in a pouch. Three samples are retained as controls with three being subjected to pressure treatment at 600 MPa for 3 min. Residual survivors are enumerated as described in the sampling plan and standard operating procedures (SOP).

3.5 Sampling plan

See Table 3.1.

Table 3.1 Experimental plan to process juices.

Product/batch	Pathogen	Container	Controls	HHP-treated
Cheeky carrot and peach				
1	*Escherichia coli* O157:H7	32 oz	3	3
	Salmonella	32 oz	3	3
	Listeria	32 oz	3	3
	Escherichia coli O157:H7	4 oz	3	3
	Salmonella	4 oz	3	3
	Listeria	4 oz	3	3
2	*Escherichia coli* O157:H7	32 oz	3	3
	Salmonella	32 oz	3	3
	Listeria	32 oz	3	3
	Escherichia coli O157:H7	4 oz	3	3
	Salmonella	4 oz	3	3
	Listeria	4 oz	3	3
3	*Escherichia coli* O157:H7	32 oz	3	3
	Salmonella	32 oz	3	3
	Listeria	32 oz	3	3
	Escherichia coli O157:H7	4 oz	3	3
	Salmonella	4 oz	3	3
	Listeria	4 oz	3	3

(*Continued*)

Table 3.1 (Continued)

Product/batch	Pathogen	Container	Controls	HHP-treated
Groovy beet and strawberry				
1	*Escherichia coli* O157:H7	32 oz	3	3
	Salmonella	32 oz	3	3
	Listeria	32 oz	3	3
	Escherichia coli O157:H7	4 oz	3	3
	Salmonella	4 oz	3	3
	Listeria	4 oz	3	3
2	*Escherichia coli* O157:H7	32 oz	3	3
	Salmonella	32 oz	3	3
	Listeria	32 oz	3	3
	Escherichia coli O157:H7	4 oz	3	3
	Salmonella	4 oz	3	3
	Listeria	4 oz	3	3
3	*Escherichia coli* O157:H7	32 oz	3	3
	Salmonella	32 oz	3	3
	Listeria	32 oz	3	3
	Escherichia coli O157:H7	4 oz	3	3
	Salmonella	4 oz	3	3
	Listeria	4 oz	3	3
Cool cucumber and pineapple				
1	*Escherichia coli* O157:H7	32 oz	3	3
	Salmonella	32 oz	3	3
	Listeria	32 oz	3	3
	Escherichia coli O157:H7	4 oz	3	3
	Salmonella	4 oz	3	3
	Listeria	4 oz	3	3
2	*Escherichia coli* O157:H7	32 oz	3	3
	Salmonella	32 oz	3	3
	Listeria	32 oz	3	3
	Escherichia coli O157:H7	4 oz	3	3
	Salmonella	4 oz	3	3
	Listeria	4 oz	3	3
3	*Escherichia coli* O157:H7	32 oz	3	3
	Salmonella	32 oz	3	3
	Listeria	32 oz	3	3
	Escherichia coli O157:H7	4 oz	3	3
	Salmonella	4 oz	3	3
	Listeria	4 oz	3	3
Total number of samples			108	108

HPP, high hydrostatic pressure

3.6 Test procedures

1. Preparation of inoculum;
2. Samples preparation and inoculation to achieve concentration upto 10^6 cfu/mL;
3. Measure product pH, Brix, initial temperature;
4. Filling and sealing of inoculated sample in PET bottles;
5. HPP treatment
   a. 600 MPa, 0–3-min hold time,
   b. 550 MPa, 0-, 3-, and 5-min hold time, and
   c. 500 MPa, 0-, 3-, and 5-min hold time;
6. Samples transportation at 0°C and microbiological analysis;
7. Data analysis and establishment of HPP critical process parameters;
8. HPP process verification; and
9. Shelf-life studies in PET bottles to verify that established HPP process will provide at least 21-day storage at 4°C. Untreated freshly made product will be used.

3.7 Source of pathogens

See Table 3.2.

Table 3.2 XXX.

	Source	Rational
Listeria monocytogenes		
7163 Serotype 1/2a	Beef	Meat isolate Serotype implicated in foodborne illness
L10540 Serotype 1/2b	Salami	Meat isolate Serotype implicated in foodborne illness
L10549 Serotype 4b	Cheese	Dairy isolate Serotype implicated in foodborne illness
L10505 Serotype 4c	Beef	Meat isolate Serotype implicated in foodborne illness
L10511 Serotype 3b	Sausage	Meat isolate

(*Continued*)

Table 3.2 (Continued)

	Source	Rational
Escherichia coli O157:H7		
ph 1	stx^1-, stx^2-*lux CDABE* (*Ph. 1*), Amp	Avirulent strain
AP 203	Beef	Meat isolate
PT120	Clinical	Previously implicated in foodborne illness outbreak
Er 92	Manure	Environmental isolate
Er50	Clinical	Previously implicated in foodborne illness outbreak
Salmonella Typhimurium DT104	Pig	Multi-drug resistant implicated in foodborne illness
Enteritidis	Poultry	Implicated in foodborne illness
Montevideo	Soil	Environmental
Heidelberg	Clinical	Multi-drug resistant implicated in foodborne illness
Newport	Tomato	Environmental
Staphylococcus aureus		
ATCC 11371	Clinical	Type strain isolated from a patient with fatal enteritis
ATCC 13565	Ham	Implicated in a foodborne illness outbreak linked to ham
PS29	Pig skin	Animal origins
NU 4	Scraper blade within a pork processing facility	Derived from meat processing environment
PH13	Pig holding area within a pork processing facility	Environmental isolate

3.8 Cultivation, enumeration, and enrichment protocols for model microbes

See Table 3.3.

Table 3.3 XXX.

Bacteria	Cultivation method	Enumeration method	Enrichment method	Confirmation
Escherichia coli O157: H7 pH1 Feng & Weagant, 2011	BHI broth 37°C for 14 h	CT-SMAC 37°C for 24 h	Buffered peptone water and incubated overnight at 37°C. Aliquots of the enriched homogenate was streaked onto CT-SMaC plates that were incubated at 37°C for 24 h	Neogen Reveal Lateral Flow Immuno-assay AOAC 2000.14
Modifications	Pyruvate was omitted from the buffered peptone water for enrichment on account of the low background microflora. No polymerase chain reaction to detect stx gene was performed with samples being directly			
Salmonella Andrews and Hammack, 2011	BHI broth 37°C for 14 h	XLT 37°C for 24 h	Buffered peptone water and incubated overnight at 37°C. Aliquots (0.1 mL) were spotted onto semisolid RV plates that were incubated at 42°C for 16 h. Positive plates (diffuse white colonies) were subcultured onto XLT plates that were incubated at 37°C for 24 h	Oxoid *Salmonella* latex test AOAC 960902
Modifications	Enrichment was performed in buffered peptone water as opposed to lactose broth. Only XLT was used as a selective agar as opposed to a combination of XLD, bismuth sulfite, and Hektoen agar. XLT has been used routinely for high hydrostatic pressure (HHP) trials with no interference from background microflora. Confirmation testing was confined to the latex agglutination tests as opposed to a range of biochemical tests. Again, this was due to the low background microflora encountered			
Listeria monocytogenes Hitchins, 2011	BHI broth 37°C for 14 h	Oxford Formula Agar 30°C for 48 h	Novel *Listeria* enrichment broth and incubated at 37°C for 24 h. Aliquots of the enriched culture was then subcultured onto Oxford formula agar and incubated for 48 h at 30°C	Oxoid *Listeria* Rapid Test AOAC 960701
Modifications	Novel enrichment broth was used in place of the primary and secondary enrichment method. Novel enrichment broth has been found to give equal performance compared to current methods with the advantage of using only a single enrichment step. Only the *Listeria* Rapid test is used for confirmation as opposed to the biochemical tests highlighted in the AOAC procedure. This was due to the low background microflora encountered			
Staphylococcus aureus	BHI broth 37°C for 14 h	Baird-Parker Agar 37°C for 24 h	None	Latex Agglutination test HPB Method MFHPB-21
Modifications	None			

3.9 HPP operation procedure

See Table 3.4.

Table 3.4 HPP test standard operating procedures (SOP).

Task #	Procedure	Methodology
1	Storing and reviving cultures	Cultures are maintained in 1-mL vials of a glycerol:nutrient broth at −80°C. The vials are removed and a loop sample streaked onto the appropriate selective agar. The plates are incubated for 1−5 days at the appropriate temperature. The cultures were checked for purity by visualization of colony morphology
2	Preparation of the inoculum	The microbes are cultivated in 50−500 mL of the appropriate broth. The cells are harvested by centrifugation (5000 × *g*, 4°C for 10 min) and resuspended in sterile saline to give an optical density at 600 nm of 0.2. The cell suspension is used directly or diluted to the appropriate level
3	Sample preparation and inoculation.	The juice sample to be tested is delivered under frozen conditions. The juice was defrosted and 800 mL was placed in a sterile container and inoculated with 0.1 mL of culture cocktail. The inoculated juice was then dispensed in 25-mL lots in plastic pouches that were vacuum sealed and placed within a further bag that is also sealed. The prepared packs are transferred to a cold box with ice packs at the bottom to ensure the temperature of the sample does not exceed 4°C. Three samples were used for each treatment condition.
4	high hydrostatic pressure (HHP) treatment	The samples are transported within the cooler to the HHP facility located approximately 40 min drive from the University. The samples are loaded into the HHP chamber and treated for the designated pressure and time combination. Upon completion of the treatment the samples are returned to the cold box. Control samples that receive no HHP treatment are transported along with the treated samples

(*Continued*)

Table 3.4 (Continued)

Task #	Procedure	Methodology
5	Microbiological analysis	The treated and control samples are returned to the laboratory and held for no longer than 16 h before being subjected to microbiological analysis. The outer bag is removed and the inner opened using sterile scissors. Aliquots (25 mL) of the sample were placed in 225 mL of saline and stomached for 60 s. A dilution series of juices were performed down to 5 dilutions for control samples (untreated) and down to 2 dilutions for HHP treated samples. Aliquots (0.1 mL) of the dilution series are then spread plated onto the appropriate agar. Nondiluted juice was also plated and incubated under the designated conditions. Colonies are counted and the log cfu/mL calculated. A colony from the plate per batch are taken for confirmation testing
7	Data analysis	The log count reduction of the HHP treatment is calculated by subtracting the log cfu/mL of HHP-treated sample from the log cfu/mL of nontreated control. The lower detection limit of the assay was 1 log cfu/25 g

3.10 Results

1. Characteristics of juice product and ingredients in formulations (Table 3.5)

   Phase 1 demonstrated that a 5-log reduction of model pathogens (*E. coli*, *Salmonella*, and *Listeria*) could be achieved through applying 600 MPa for 3 min at 11°C in three types of high acid juices tested.

   The results indicated that the HPP treatment was beyond that expected to achieve the food safety objective of 5 log cfu according to juice hazard analytical critical control points (HACCP). The 600 MPa is considered as an industry standard practice for high acid juice HPP treatment. Reducing the applied pressure and treatment time offers advantages on efficiency, better juice quality in addition to less strain on the HPP unit and higher daily output through HPP processing cycle reduction.

Table 3.5 pH and Brix of tested juices.

Juice type	Ingredients	pH	Brix
Cheeky carrot	Carrot, apple, orange, inulin, lime	3.96	4.5
Groovy beet	Apples, cucumbers, beet, strawberry, inulin	3.60	5.0
Cool cucumber	Apples, cucumbers, spinach, inulin	3.70	4.5

Note: The pH measurement showed that tested juices belong to high acid juice categories with pH < 4.0.

2. Microbial challenge studies at 600 MPa for 3 min (Tables 3.6–3.8)

Table 3.6 *Escherichia coli* O157:H7 inactivation in juice products.

Juice type	Sample	Log reduction count	
		Log cfu/mL	Average ± SD
Groovy beet	I	6.51	6.57 ± 0.06
	II	6.61	
	III	6.58	
Cheeky carrot	I	6.66	6.66 ± 0.07
	II	6.72	
	III	6.58	
Cool Cucumber	I	7.12	7.13 ± 0.01
	II	7.14	
	III	7.13	

Table 3.7 *Salmonella* inactivation in juice products.

Juice type	Sample	Log reduction count	
		Log cfu/mL	Average ± SD
Groovy beet	I	6.78	6.92 ± 0.18
	II	7.12	
	III	6.86	
Cheeky carrot	I	6.99	6.86 ± 0.14
	II	6.71	
	III	6.88	
Cool cucumber	I	7.10	7.11 ± 0.06
	II	7.06	
	III	7.18	

3. The effect of HPP process pressure and treatment time on microbial inactivation (Table 3.9)

For *E. coli* O157:H7, the extent of inactivation was dependent on the pressure treatment and juice type. Specifically, cheeky carrot supported

Table 3.8 *Listeria monocytogenes* inactivation in juice products.

Juice type	Sample	Log reduction count	
		Log cfu/mL	Average ± SD
Groovy beet	I	5.97	5.95 ± 0.05
	II	5.89	
	III	5.98	
Cheeky carrot	I	6.08	6.12 ± 0.13
	II	6.27	
	III	6.03	
Cool cucumber	I	6.03	6.09 ± 0.06
	II	6.10	
	III	6.14	

Table 3.9 *Escherichia coli*, *Salmonella*, and *Listeria* inactivation in juices at 550–600 MPa.

Pathogen/juice	Log count reduction (positive by enrichment/total tested)				
	550 MPa		600 MPa		
	60 s	120 s	60 s	120 s	180 s
Escherichia coli O157:H7					
Groovy beet	3.17 ± 0.14	3.74 ± 0.07	4.28 ± 0.02	4.34 ± 0.09	6.24 ± 0.58 (1/3)
Cheeky carrot	4.43 ± 0.13	5.08 ± 0.12	6.14 ± 0.24	6.07 ± 0.11	6.66 (0/3)
Cool cucumber	3.58 ± 0.06	3.66 ± 0.17	4.63 ± 0.08	4.66 ± 0.04	5.25 ± 0.25
Salmonella					
Groovy beet	6.92 (0/3)	6.92 (0/3)	6.92 (0/3)	6.92 (0/3)	6.92 (0/3)
Cheeky carrot	6.86 (0/3)	6.86 (0/3)	6.86 (0/3)	6.86 (0/3)	6.86 (0/3)
Cool cucumber	7.11 (0/3)	7.11 (0/3)	7.11 (0/3)	7.11 (0/3)	7.11 (0/3)
Listeria monocytogenes					
Groovy beet	4.36 ± 0.11	4.43 ± 0.03	4.84 ± 0.19	5.07 ± 0.16	5.86 ± 0.02
Cheeky carrot	5.12 (3/3)	5.12 (3/3)	5.12 (3/3)	5.12 (3/3)	5.45 ± 0.58 (2/3)
Cool cucumber	5.09 (3/3)	5.09 (3/3)	5.09 (3/3)	5.76 + 0.58 (2/3)	6.09 (0/3)

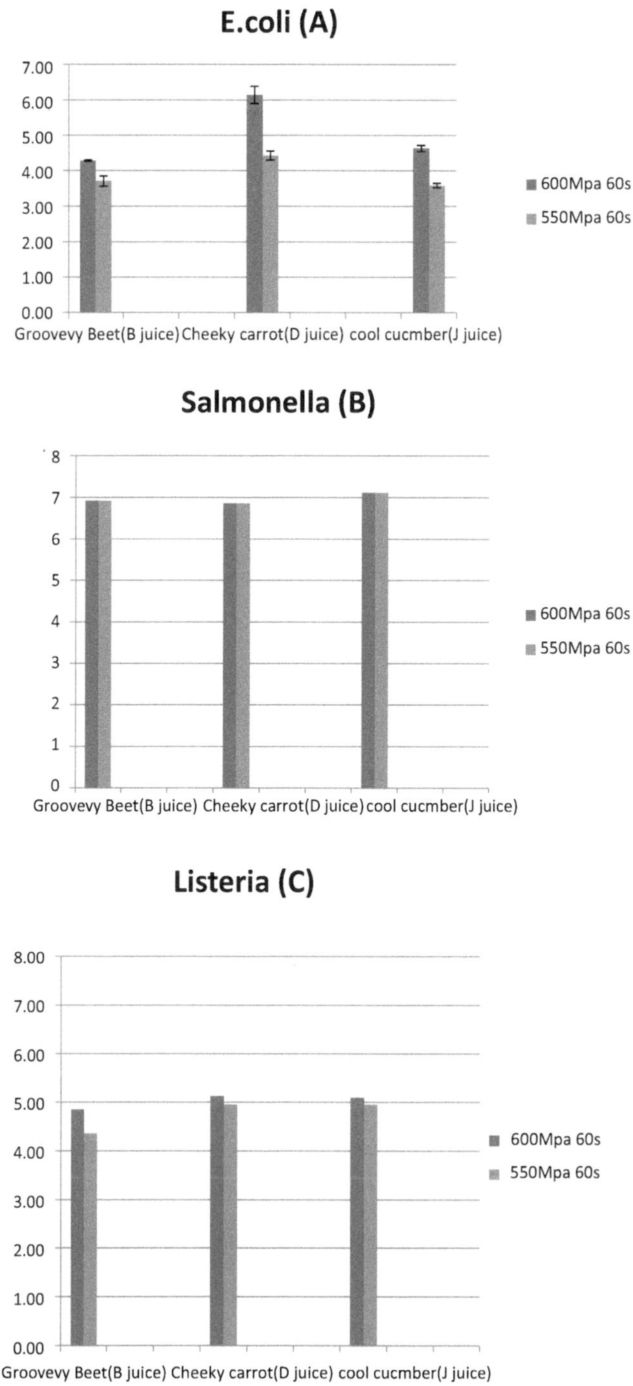

Figure 3.1 Effect of pressure of 550–600 MPA on the inactivation of *Escherichia coli* (A), *Salmonella* (B), and *Listeria* (C).

the highest level of *E. coli* O157:H7 inactivation with 550 MPa for 120 s or greater pressure/time achieving >5 log cfu reduction. For the other juices tested, a pressure of 600 MPa for 180 s was required to achieve the target log reduction of the pathogen.

A >5 log cfu reduction in *Salmonella* was achieved at all the pressure and times tested.

A >5 log cfu reduction of *L. monocytogenes* could be achieved at the pressure and times tested with cheeky carrot and cool cucumber.

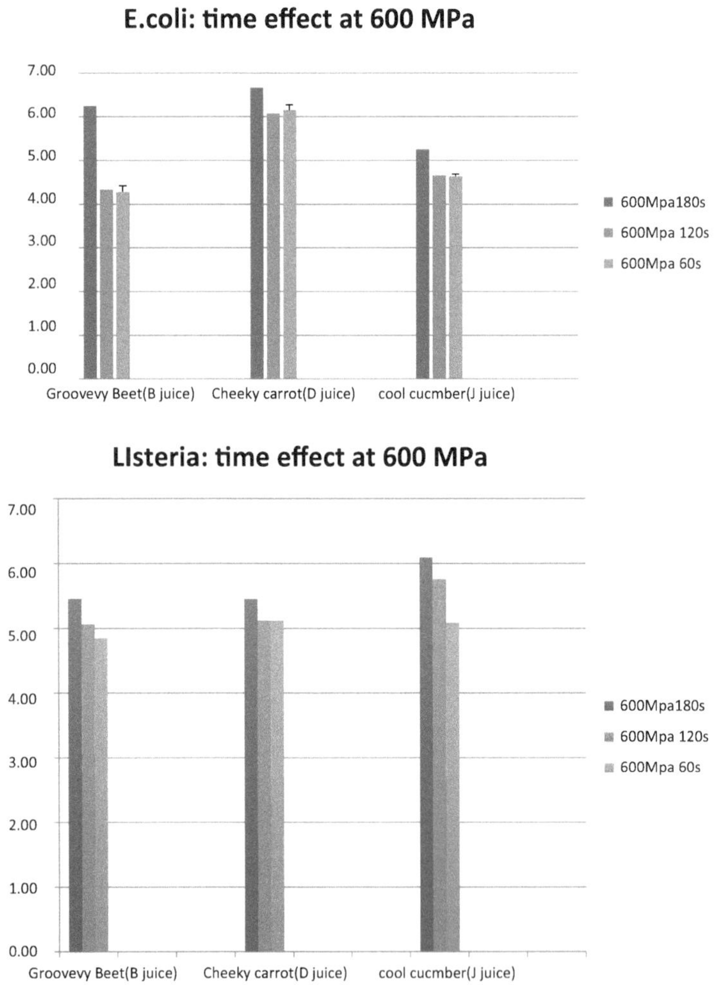

Figure 3.2 Effect of treatment time at 600 MPA on the inactivation of *Escherichia coli* and *Listeria*.

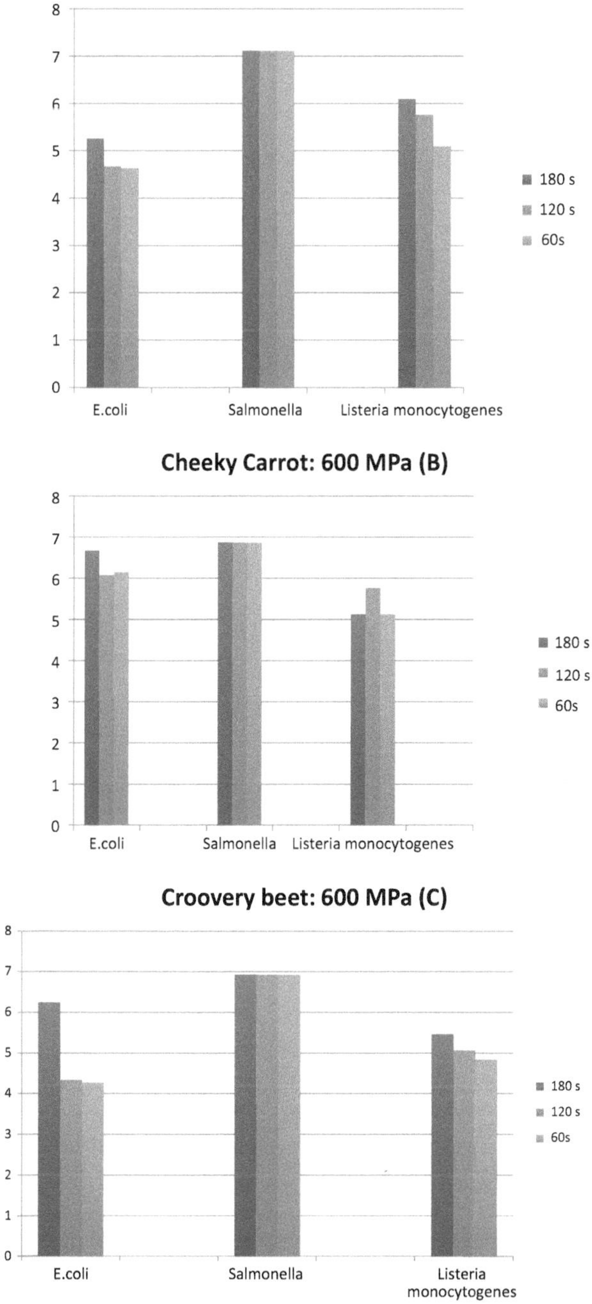

Figure 3.3 Effect of juice composition at 600 MPA on the inactivation of *Escherichia coli (A)*, *Salmonella (B)*, and *Listeria (C)*, Y axis - Log reduction.

However, the pathogen was more barotolerant in groovy beet with treatments 600 MPa and time >60 s required to attain the target log reduction of 5 log cfu.

The effects of process pressure, treatment time, and juice composition on inactivation of pathogens are shown in Figs. 3.1–3.3.

3.11 Conclusion

The extent of pathogen inactivation with respect to *E. coli* O157:H7 and *L. monocytogenes* was found to be dependent on the processing conditions along with the juice type. Interestingly, *E. coli* O157:H7 exhibited higher barolerance in the beet- and cucumber-based beverage whereas *Listeria* was resistant in only the beet. The results indicate that juice formulation has a role in defining the pressure resistance of pathogens, although which constituents provide protection to HPP remains unclear.

References

Andrews, W.H., Hammack, T., 2011. *Salmonella.* <http://www.fda.gov/Food/ScienceResearch/LaboratoryMethods/BacteriologicalAnalyticalManualBAM/ucm070149.htm> [accessed September 1, 2021].

Feng, P., Weagant, S.D., 2011. Diarrheagenic *Escherichia coli*. BAM Ch 4. Available from: http://www.fda.gov/Food/ScienceResearch/LaboratoryMethods/BacteriologicalAnalyticalManualBAM/ucm070080.htm [accessed September 1, 2021].

Hitchins, A.D., 2011. Detection and enumeration of Listeria monocytogenes in foods. BAM Chp 10. Available from: http://www.fda.gov/Food/ScienceResearch/LaboratoryMethods/BacteriologicalAnalyticalManualBAM/ucm071400.htm [accessed September 1, 2021].

CHAPTER 4

Validation of light-based processes

Tatiana Koutchma
Novel Food Processing Technologies, Agriculture and Agri-Food Canada, Guelph, ON, Canada

4.1 History and introduction to fundamentals of ultraviolet light and pulsed light technology for food applications

Ultraviolet (UV) technology is a physical intervention towards improved hygiene, safety control, and extended shelf-life (ESL) of foods and beverages in the food industry. With the growth of fresh, natural, and ready-to-eat (RTE) markets and new disinfection methods, continuous germicidal ultraviolet-C (UVC) light at 253.7 nm and other light technologies are used as dry, nonthermal, nonionizing, and nonchemical food safety and preservation techniques, which can be applied for the treatment of the surfaces, liquid foods, beverages, and ingredients. The discovery of UV inactivation of the chlorine-resistant parasites *Cryptosporidium parvum* and *Giardia* sp. has catalyzed for the use of UV light in the drinking water industry and treatment of waste and processing water. Despite the long history of use for drinking and wastewater treatment, air disinfection, and surface decontamination, there is still its limited use in food supply chains. Based on engineering advances and new scientific data, light technologies in continuous and pulsed modes [UV, pulsed UV (PUV), and pulsed light (PL)] offer promise of improved microbiological and chemical safety of solid and liquid foods, juices, and beverages while extending their shelf-life, enhancing functionality of whole and fresh cut produce and has a potential for obtaining premium quality products that can lead to the faster commercialization. Reports are available that application of UV light can also improve the toxicological safety of foods of plant origin through its ability to reduce the levels of toxins, such as patulin mycotoxin or mycotoxins, on grain and corn products.

Sanitation, decontamination, disinfection, and advanced oxidation with UV and PL are the versatile, environmental-friendly technology, which can be used in the food production and storage facilities to reduce

Validation of Food Preservation Processes based on Novel Technologies
DOI: https://doi.org/10.1016/B978-0-12-815888-3.00005-1

microbial contamination of food contact surfaces and consequently to improve the safety of finished products. In addition, with the growth of fresh, natural, and RTE markets, UV and other light technologies are used as a dry food safety and preservation technique when applied for the treatment of the food products' surface due to their unique bacterial effects and minimal impact on food quality.

A number of light-based techniques have been explored for food applications due to unique properties of light to inactivate microflora at the air–surface interface and minimal impact on food quality. The investigated light technologies include continuous UV light and light-emitting diodes (LEDs), PUV light and PL. Each technology utilizes a different type of light that is generated in various forms and demonstrates different disinfection efficacies depending on the applications.

UV light is known as a portion of the electromagnetic spectrum between 100 and 400 nm corresponding to photon energies from 124 to 3 eV. The wavelength of UV light is shorter than the violet end of the visible spectrum, but longer than X-rays. Because of low inherent energy of photons, the penetration of UV photons is low in comparison with ionizing radiation, and UV light therefore belongs to the group of nonionizing radiation. UV light range is further divided into UVA, -B, and -C subdiapasons, which are characterized by various wavelengths and energy levels of photons and thus differ in their effects. Most people are familiar with UVB and UVA from sunlight as causing suntan and sunburn. UVC light in the bactericidal range from 200 to 280 nm effectively inactivates bacterial, viral, and protozoan microorganisms in air, water, surfaces, and foods.

PL has a wide range of wavelengths, from 170 to 1000 nm, and often combines visible, UV, and infrared (IR) radiation at energies of 0.01–50 J/cm^2. Varying in terms of wavelength, frequency, and energy, the emitted broad spectrum radiation of xenon lamps ranges from approximately 180 to 1100 nm and encompasses next to the IR (700–1100 nm), visible light (VIS, 400–700 nm), and the fraction of UV light (180–400 nm), each contributing 30%, 45%, and 25% to the whole spectrum, respectively. However, the UVC part of the PL spectrum is the most important for microbial inactivation.

PL is delivered in several flashes of light per second, allowing fast throughput of product and low energy usage. The intensity of PL is about 20,000 times the intensity of continuous UV light and provides some practical advantages over continuous UV light sources in the situations where rapid treatment is required.

The efficacy of UV light sources against different groups of pathogenic and spoilage organisms varies in different matrices, such as air, water, surfaces, or fluid foods. The goal of this chapter is to discuss validation principles of light-based processes for the treatment of surfaces and liquid products and provide guidelines for technology implementation.

4.2 Continuous UV, pulsed, and LED light sources

The source characteristics, such as wavelength, that defines the levels of photon energy, the light source intensity, and exposure time will define process lethality.

There is a large number of manufacturers of UV equipment suitable for food processing facilities and rooms using various UV sources including low-pressure mercury (LPM) and low-pressure amalgam (LPA) lamps and pulsed lamps that are typically supplied with automated, online wipers that keep lamps free from fouling. The market of UV disinfection equipment is expected to grow to $1.1 billion in 2013 and $2.1 billion in 2018 for a compound annual growth rate of 13.1%.

Lamps' manufacturers use the following characteristics to compare electrical and UV efficiency of UV sources: total power input, UVC efficiency, irradiance, and irradiance at a given distance and lamps lifetime.

- *Total power input* (W) depends on voltage and electric current.
- *UVC efficiency* of the lamp is evaluated based on the measured wattage of UV output in the spectral range of interest versus the total wattage input to the lamp.
- *Irradiance* is the amount of flux incident on upon a predefined surface area and most commonly expressed in mW/cm^2 or W/m^2.
- UV irradiance at a given distance from the lamp(s) surface (10 or 20 cm).

Continuous LPM and LPA UVC lamps that emit light photons at 253.7 nm and xenon PL lamps with emission spectrum from 200 to 1000 nm are currently the dominant sources of radiation for microorganism control since they have been approved by the US Food and Drug Administration (FDA) CFR 21 179.39 as described in detail in Section 4.3.

The light sources' disinfection efficiency depends on the germicidal range and target organism. For example, LPM lamps emit UVC light at 254 nm near the peak of DNA and RNA absorbance (260–280 nm) to inactivate microorganisms. Polychromatic broader wavelength spectrum affects the DNA and RNA plus other biological molecules such as proteins and enzymes, enabling greater inactivation impact. A wide germicidal wavelength attacks microorganisms on several targets, inflicts damage to the microorganisms' essential components, and inhibits the microorganisms' mutation repair mechanism. For example: the absorbance spectra of proteins show a maximum peak at 280 nm, whereas the peptide bond in proteins displays a significant absorbance below 240 nm. Another example is *Cryptosporidium* and *Bacillus subtilis* spores, which are inactivated most efficiently at 270–271 nm, beyond the scope of emitting spectra of low-pressure lamps.

LPM and LPA lamps at 253.7 nm are mainly adapted by the food industry as sources of continuous monochromatic UVC light. LPM and LPA lamps are readily available at comparatively low cost. Standard quartz LPMs and LPAs are used across the whole range of surface applications for dry disinfection and sanitation processes. LPM and LPA lamps have a reasonably long lifetime of up to 18,000 h. The optimum pressure of mercury is about 1 Pa, and the optimum temperature of LPM lamps is around 40°C. At wavelengths below 200 nm, such as 185 nm, ozone is produced from oxygen and organic compounds can be oxidized. For the food industry, the production of ozone is not desired, not only because it has no germicidal effect on microorganisms, but it can also cause side reactions with organic components in the food. If the 185 nm wavelength is unnecessary, for example, if no oxidation is required, special lamps using doped quartz can be used to completely filter out this wavelength. UV irradiance of LPM lamps is typically up to 10 mW/cm^2. LPM lamps are easy to install and operate at maximum temperature up to 40°C and have comparatively low cost. The energy losses are mainly in the form of heat.

A concern over the impact of mercury release from lamps into the plant environment encouraged the research of mercury-free lamps. Instead of mercury, amalgam can be used in LPA lamps, resulting in an extremely long life while ensuring high power outlet. Up to ten times the UV power density of LPM can be achieved while they can be used at higher ambient temperatures of up to 90°C. The UVC irradiance of LPA lamps is insensitive to temperature fluctuations. The amalgam lamps do not show the transmission loss of quartz glass associated with LPM lamps and therefore produce a constant inactivation action over the total operating life of the lamp. Compared to LPM lamps with an UVC output of approximately 40%, up to 90% of the UVC output power is delivered with the use of amalgam lamps. Depending on the type of quartz used, ozone-free or ozone-generating lamps are available.

Commercial UVC systems are developed and can be readily integrated in food facilities at various points of production.

Xenon lamps are commercial sources of PL that are more environmentally friendly than UV lamps because they do not use mercury. However, they require air or water cooling for their operation. Xenon flash lamps have an emission spectrum ranging from UV to IR light. The UVC part of the spectrum is the most important for microbial inactivation.

Future tunable monochromatic and polychromatic sources would allow users to adjust the wavelength to specifically match the needs of the application. The commercial UV-LED can emit in three diapasons of UVA, UVB, and UVC and have the following advantages: small size, long lifetime up to 10,000 h for UVB and UVA range, energy-efficient, easy control of emission, and no production of mercury waste and glass. The germicidal effects against several bacteria or fungi have been demonstrated and reported along with the first commercial applications for disinfections of hospital rooms, water, and various point of use applications.

The LED is a new technological solution with enormous potential for food processing operations. There has been comparatively limited number of studies involving UV LEDs for food treatments in the UVC range likely due to their high cost and comparatively lower irradiance. So far, the majority of studies exploring effects of UV LEDs against food pathogens and fungal spoilage organisms were conducted in model systems. So, the study by Green et al. (2018) reported the lethal effectiveness of UV LEDs emitting in 255–365 nm range against common food-borne pathogens, such as *Escherichia coli*, *Salmonella*, and *Listeria monocytogenes*. The UVC LEDs emitting light in the approximate range of 265 – 280 nm were found to have the greatest efficacy in terms of bacterial inactivation and overall power output. Also, the tested pathogenic organisms exhibited peak UV sensitivity at special wavelengths. Among three tested bacteria, *Salmonella enterica diarizonae* and *E. coli* O157:H7 exhibited the highest overall UV sensitivities, which occurred at 259 nm. In contrast, *L. monocytogenes* peaked in UV sensitivity at 268 nm. The inactivation data between 253.7 and 370 nm were used to construct UV inactivation spectra for tested food pathogenic and nonpathogenic strains.

To displace the incumbent mercury-based lamp technology, UV LEDs require further development to improve device efficiency and lifetime. The biggest challenge with UVC LEDs shorter than 265 nm is that they are still weak light emitters and have relatively short lifetime. As UV

LEDs' efficiency improves at lower cost of their production, lifetime increases and engineering design principles better develop, LEDs fixtures will begin to replace the existing UVC lamps market as a cost-effective, environmentally friendly alternative to save energy, improve safety, and shelf-life of produced foods from field to fork. It is very likely that in the near future many applications that today use mercury lamps will be soon carried out by UV LEDs. Comparative summary of basic technical characteristics and efficiency of commercial continuous UVC and PL sources is presented in Table 4.1.

From this summary it is evident that no single lamp technology that will represent the best source for all applications. However, situation-specific requirements, such as required production rate, efficiency, temperature, and others, might dictate a clear advantage for a given light technology.

4.3 International regulatory requirements

4.3.1 US Food and Drug Administration: food additive approach

According to CFR 21 part 179, the US FDA considers UV light as radiation, along with ionizing radiation, radiofrequency radiation, PL and carbon dioxide laser and a source of radiation used to treat food, as a food additive (US FDA, 2001). The additive is not literally added to food. Instead, a source of radiation is used to process or treat food such that, analogous to other food processes and its use, it can affect the characteristics of the food. In assessing the safety of foods treated with all forms of radiation, the agency considers microbial efficacy, changes in chemical composition of the food that may be induced by the proposed treatment, including any potential changes in nutrient levels (US FDA, 2001)

4.3.1.1 Continuous UVC light

4.3.1.1.1 Juice products

In 2001 the US FDA amended the food additive regulations to provide for the safe use of UV radiation at a wavelength of 253.7 nm to reduce human pathogens and other microorganisms in juice products. It was determined that the amount of UV irradiation necessary for human pathogen reduction would depend on the type of juice, the initial microbial load and the design of the irradiation system (e.g., flow rate, number of lamps and time exposed to irradiation). Therefore the FDA did not

Table 4.1 Summary of basic parameters of continuous UVC and pulsed light and UV sources.

UV source	Electrical efficiency (%)	UVC efficiency (%)	UV intensity (W/cm^2)	Lamp surface temperature (°C)	Treatment timescale	Lifetime (months)	Output spectrum
LPM UVC	50	30–38	0.01	40	Seconds to minutes	18–24	Monochromatic 253.7 nm
MPM	15–30	10–12	600	400–1000	Seconds	0.5	Polychromatic (200–600 nm)
PUV					Milliseconds to seconds	1	Polychromatic
	15–20	17	30,000	–		NA	Polychromatic 200–400 nm
UV-LED	60	5–8	0.1–0.001	Ambient	Hours to minutes	> 10	Monochromatic
UVA, UVB, UVC							254–400 nm

LPM, Low-pressure mercury; *MPM*, medium-pressure mercury; *PUV*, pulsed UV; *UV*, ultraviolet.

specify a minimum or maximum UV dose by regulation but concluded that this should be achieved for individual usage situations in a manner consistent with good manufacturing practice.

In addition, the FDA expects that the maximum dose applied to the juice will be economically self-limiting due to the costs associated with UV irradiation. Also, the levels of UV irradiation applied to the juice will be limited by the possible alterations in quality, nutritional, and organoleptic characteristics of the juice (i.e., changes in taste or color) after UV irradiation and changes that may result in decreased consumer acceptance.

4.3.1.1.2 Surface and potable water

Other approved applications of UV radiation for the processing and treatment of foods are surface microorganisms control and sterilization of potable water. Low-pressure lamps operating at a wavelength of 253.7 nm are approved as radiation sources with the limitations in terms of ozone production. The essential treatment conditions in terms of intensity, flow patterns, and product requirements are outlined in the US FDA CFR 21 179.39 (Table 4.2).

Table 4.2 US FDA CFR 21 179.39—UV for treatment and processing foods.

Irradiated food	Limitations	Use
Food and food products	Without ozone production: high-fat content food irradiated in vacuum or in an inert atmosphere; intensity of radiation, 1 W (of 2537 Å radiation) per 5–10 ft^{2}[a]	Surface microorganism control
Potable water	Without ozone production; coefficient of absorption, 0.19/cm or less; flow rate, 100 gal/h/W of 2537 A radiation; water depth, 1 cm or less; lamp-operating temperature, 36°C–46°C	Sterilization of water used in food production
Juice products	Turbulent flow-through tubes with a minimum Reynolds number of 2200	Reduction of human pathogens and other microorganisms

FDA, Food and Drug Administration; *UV*, ultraviolet.
[a]UV intensity corresponds to 0.21–0.107 mW/cm^2.

4.3.1.1.3 UV-pasteurized water for dairy industry

In 2009 FDA Pasteurized Milk Ordinance (PMO, 2009) approved UV disinfection to create pasteurized equivalent water for use in cleaning and rinsing applications. The regulation states that UV light shall be applied so that the entire volume of water receives at least the following dose when used as pasteurized water.

- Low-pressure UV at 254 nm at a UV dose of 186 mJ/cm^2 or a 4-log adenovirus equivalent;
- Medium-pressure UV at 120 mJ/cm^2 or a 4-log adenovirus equivalent.

Other requirements include various control elements to ensure the all water volume receives the minimum dose of UV light, real-time monitoring to ensure that the dose is consistent, and cleaning protocols to ensure the system always works at peak performance.

4.3.1.1.4 Baking yeasts

The US FDA has approved UV light treatment of baker's yeast to convert endogenous ergosterol to vitamin D2. The FDA concluded that the use of UV light-treated baker's yeast does not pose a safety concern, since the UV light treatment has been shown not to produce any new components of toxicological concern that could be introduced into the diet (U.S. Food and Drug Administration, 2012, 21 CFR Part 172, 381). Vitamin D2 baker's yeast may be used safely in foods as a source of vitamin D2 and as a leavening agent.

4.3.1.2 Pulsed UV light in the production, processing, and handling of food

Code 21CFR179.41, issued by the FDA, Department of Health and Human Services, approves the use of PUV light in the production, processing, and handling of food (US FDA, 21CFR179.41). PL may be safely used for the treatment of foods under the following conditions:

1. The radiation sources consist of xenon flash lamps designed to emit broadband radiation consisting of wavelengths covering the range of 200–1000 nm, and operated so that the pulse duration is no longer than 2 μs.
2. The treatment is used for surface microorganism control.
3. Foods treated with PL shall receive the minimum treatment reasonably required to accomplish the intended technical effect.
4. The total cumulative treatment shall not exceed an energy flow of 12.0 J/cm^2.

4.3.2 Novel food regulations

In six countries, UV-treated food products belong to the category of Novel Foods, including EU, Great Britain, Canada, Australia, New Zealand, and China. Novel Foods and ingredients are regulated in a varying manner in each country, with the majority of systems based on a risk or safety assessment review model, with most also requiring notification and approval. In general, foods that result from a process that has not been previously used for food production are considered as Novel Foods. Novel Foods are also products that do not have a long history of safe use as a food.

4.3.2.1 Canada

4.3.2.1.1 Apple juice and cider

Foods and feed that result from a UV process that has not been previously used for their production are considered as Novel Foods and Novel Feed. Companies are required to submit detailed scientific data for review and approval by Health Canada or Canadian Food Inspection Agency before such products can be sold. In 2004 Health Canada determined that there are no safety concerns, and it has no objection regarding the sale of UV-treated apple cider using commercial CiderSure 3500 unit (FPE, Rochester, NY) to achieve a reduction in the microbial load of apple juice and cider products (Health Canada, 2004).

4.3.2.2 European Union

In assessing the safety of foods treated with all forms of radiation, the agency considers microbial efficacy, changes in chemical composition of the food that may be induced by the proposed treatment, including any potential changes in nutrient levels. The legal status of light treatments in the European Union (EU) has a different approach since the legislation is not technology oriented but food and food ingredient oriented. Light technologies would fall in the scope of regulation 258/97 on novel foods and novel food ingredients, article 1, item f (European Union, 1997). The approval process is based on the risk assessment approach that includes microbiological, chemical, and toxicological safety, possible changes in composition and nutrition content, allergenicity, and dietary exposure.

Novel Food Regulation that came into effect in 2018 specifies that a food product should be considered a novel food when it results from a production process not used within the Union before May 15, 1997, and that gives rise to significant changes in the composition or structure of a food affecting its nutritional value, metabolism, or level of undesirable substances (Regulation EU, 2015).

4.3.2.2.1 Milk

In January 2016, the European Food Safety Authority (EFSA) approved the usage of UV light as a milk postpasteurization treatment due to the nutritional value of such a treatment (EFSA, 2016). The novel food is cow's milk (whole, semiskimmed, or skimmed) to which a treatment with UV radiation at 253.7 nm is applied after pasteurization to extend the shelf-life of the milk and to increase vitamin D3 concentrations by conversion of 7-dehydrocholesterol to vitamin D3. The EFSA panel concluded that the novel food, UV-treated milk, is safe under the intended conditions of use as specified by the applicant.

4.3.2.2.2 Bread

In 2014 the EFSA (2018) approved application of the equipment manufacturer for the UV treatment of bread. The permission was issued for the sale of yeast-leavened bread and rolls that can be UV-treated after baking to convert ergosterol to vitamin D2 (ergocalciferol) at the maximum level of 3 μg vitamin D2 per 100 g. The permitted UV irradiation process is within the wavelength of 240–315 nm for the maximum of 5 s with a UV dose of 10–50 mJ/cm^2.

Other authorized UV-treated products that are included in the new EU union list include mushrooms and baker's yeast (EFSA, 2018).

4.4 Validation of UV light technologies for surface treatment

The term "surface" in food processing includes the surface of the product, package, transport of product, and tools of the equipment that touch the product. The control of the surface contamination through the exposure to continuous UV or PL can be applied in prepackaging and postpackaging operations and may include surface of incoming packaging, belts or trays, unpacked food and packaged food, and surface of tools. When using UV light for surface treatments, the application and environment become as critical as the product characteristics.

As a prepackaging control measure, UV or PL treatment of packaging in filling plants, for example, lids, cups, and packaging foils for drinks and beverages help to eliminate microbial contamination and thus extend shelf-life of products. When continuous UV or PUV is applied as postpackaging treatment for finished packaged products, the considerations about transparency are critical to the packaging materials. UV and PL light can also be used as a final intervention step for the decontamination of the surface of the packaged food in the clean production areas that will be placed directly into the package with RTE products, such as packets of sauces, jams, and other items.

It is well known that preventing cross-contamination of surfaces is a key factor in reducing the risk of food-borne illnesses and thus improving food plants safety. Light-based approaches were tested to reduce the contamination of nonfood contact surfaces, food contact surfaces, and foods at different production points and points of use applications. The investigated light technologies included continuous UV light in germicidal UVC range and LEDs, PUV light, and PL.

As indicated earlier, a variety of UV sources are commercially available or currently under development that can be applied for specific purposes at the food plant for surface treatment whereas LPM lamps are currently the dominant sources of radiation for the processing and treatment of foods since they were approved by the US FDA CFR 21 179.39 (Table 4.2) (U.S. Food and Drug Administration, 2001). LPM lamps are allowed for surface organisms control without any ozone production in vacuum or inert gas atmosphere. The intensity of UVC light at 253.7 nm is established 1 W per area of 5–10 ft^2. PL may be safely used when the radiation sources consist of xenon flash lamps emitting radiation in the range of 200–1000 nm and with the pulse duration is no longer than 2 ms; and the total cumulative treatment shall not exceed 12.0 J/cm^2.

This section deals with the applications of light technologies as a complementary measure to control spreading bacteria and viruses through contact surfaces.

4.4.1 UV disinfection efficiency of surfaces and limiting factors

There are a number of essential characteristics of light source, surface parameters, and UV resistance of microorganisms that define the efficiency of UV disinfection.

First, the light sources' efficiency depends on the germicidal range and UV resistance of target organism. The reported data demonstrated that there is a difference in UV resistance of different groups of microorganisms that follows the order: mold spores > viruses > bacterial spores > yeasts > vegetative bacteria.

It was reported that Gram-positive bacteria are generally more resistant to UV light than Gram-negative bacteria. Beauchamp and Lacroix (2012) reported that *L. monocytogenes* produced 35% fewer cyclobutane pyrimidine dimers CPDs and 10% fewer pyrimidine 6–4 pyrimidone photoproducts than *E. coli* at a UVC dose of 3 J/cm^2. This low production of UV photoproducts indicates a greater resistance for Gram-positive bacteria. Also, after *L. monocytogenes*, *Salmonella* is more resistant to UV than *E. coli*. In addition, efficacy of UV sources against different groups of pathogenic and a spoilage organism varies in different matrices, such as air, water, surfaces, or fluid foods. Adenovirus is considered as the most UV-resistant organism in water treatment, whereas the highest UV resistance in air and food matrices characterizes bacterial and mold spores.

As discussed earlier, LPM sources emit UV light at 254 nm near the peak of DNA and RNA absorbance (280 nm) to inactivate microorganisms. A wide germicidal wavelength attacks microorganisms on several targets, inflicts damage to the microorganisms' essential components, and inhibits the microorganisms' mutation repair mechanism. This means that characteristics of UV source such as wavelength or the levels of UV light photons energy, the number of photons or UV intensity, and exposure time will define UV process lethality

Other factors interfering with UV light and limiting its efficiency are distance between the light sources and treated surface. Light photons emitted from the source propagate away interacting with the materials through absorption, reflection, refraction, and scattering. The correct positioning of light source and distance from the treated product is critical to maximize the efficacy of UV and pulsed treatments. Any obstruction to the path of the light, such as dust, shadowing, or clumping of bacteria, can reduce efficacy.

The characteristics of the surface are another limiting factor because the effectiveness of light treatments is lower on a rough surface than on a smooth one. Among critical product parameters, temperature, chemical composition, and physical structure may affect UV or PL effectiveness to control microorganisms.

4.4.2 Advantages and disadvantages of high-intensity pulses

The short pulse width and high doses of the PL source may provide some practical advantages over UV sources in those situations where rapid disinfection is required.

Other advantages of PL treatment are the lack of residual compounds, and the absence of applying chemicals that can cause ecological problems and/or are potentially harmful to humans. Xenon flash lamps are also more environment friendly than continuous UV lamps because they do not use mercury.

Sample heating is perhaps the most important limiting factor of PL for practical applications. Heat can originate from the absorption of IR part of PL by the food or by lamp heating. Another disadvantage of PL treatment is the possibility of shadowing occurring when microorganisms readily absorb the rays.

4.4.2.1 Factors affecting interaction between high-intensity pulses and materials

The factors determining the interaction of PL exposure with foods are in some extent similar to UV light treatments. The critical factor affecting PL treatments is fluence incident on the sample. The energy emitted by the flash lamp is different from the energy incident on the sample. Factors, such as distance from light source to target and propagation vehicle (air, water, or other fluids, dust particles), affect the level of energy than ultimately reaches the target. The inactivation efficacy of PL is higher when treated products are closer to the lamp. Food composition also affects the efficacy of the decontamination by PL (Gomez-Lopez et al., 2007). High protein and fat containing food products have little potential to be efficiently treated by PL. Vegetables, on the other hand, could therefore be suitable for PL treatment.

4.4.3 Establishing steps and validation of surface treatment process

As a first step, the required design UV dose D (UV_design) or design reduction equivalent dose (RED) and process/product critical parameters have to be established to consistently achieve the target logarithmic inactivation or *specific logarithmic reduction* (SLR) of the target pathogen of concern on the surface of the material. The design UV dose in mJ/cm^2 on the surface of the material can be determined from Eq. (4.1) as a product of the decimal reduction dose (DRD) or D_{10uv} value (mJ/cm^2) and SLR.

$$D_{UVdesign} = \text{SLR} \times D_{10uv} \tag{4.1}$$

To calculate the design RED dose (mJ/cm^2), the DRD (D_{10uv} value) or dose to achieve 1-log reduction of the target pathogen has to be known. Also, the determination of D_{10uv} values or inactivation rate constants can be challenging because UV microbial inactivation on surfaces more often does not follow linear or first-order reaction and tailing effect has been reported. In this case, nonlinear tail model can be used to determine inactivation rate parameters. Moreover, D_{10uv} value depends on the surface characteristics and can increase dramatically on rough surfaces. Similar to established practices in UV water treatment, D_{10uv} value can be measured using collimated beam (CB) bench top using where incident UV intensity on the product surface can be controlled.

Next, the applied or operational continuous UV or PL dose (mJ/cm^2) that is generated by an incident light emitted by the source on the surface of the product should be determined from Eq. (4.2).

$$\text{UV}_{\text{dose applied}} = I_{uv} \times t \tag{4.2}$$

where I_{uv} is the UV or pulse light intensity on the surface of the product (mW/cm^2) and t is the exposure time in seconds. UV intensity is the light power incident on a unit area perpendicular to the direction of propagation. UV or PL intensity describes the magnitude of light measured with a radiometer or UV sensor.

Applied dose reflects the energy emission from the light source and it is independent of the material to be irradiated. However, I_{uv} is a function of the lamp type, output power (W), and number of lamps, age, sleeves material, geometry, and distance from the surface.

To meet the process performance objectives to achieve required SLR of organism of concern, the operational UV dose has to be at least equal or higher than the design UV dose (Eq. 4.3).

$$\text{UV dose applied} \geq \text{UV design} \tag{4.3}$$

Knowledge of the design dose allows evaluating treatment time or selecting type of UV or PL lamps that will suit the food plant environment. The lamp type, output power, length, and the number of lamps will depend on the dimensions of the total area where the product will be placed and treated. To evaluate the lamp output, light source efficiency has to be considered in this calculation. For example, LPM UV lamps convert approximately 38% of their input power into UVC Output power. This means, if the required UVC is 84 W, the total lamp power will be 84 W/0.38 = 221 W. The length of the UV lamp is another critical factor when establishing the dimensions of the UV exposure chamber. To design or to select commercially available UV system, the treatment surface area and the length of the lamp should be known.

The primary objective of process validation is to demonstrate that light-based process will consistently deliver the required performance or safety objective measured as SLR = log (N_0/N) where N_0 is the microbial count level with no UV exposure and N is the count level with UV exposure over time across pilot and later manufacturing environment with various product batches and operating ranges.

The overall validation testing should be carried out at an established validation facility or by a qualified third-party professional following a standard protocol. Currently, there is no any standard protocol for validating UV disinfection units for surface disinfection. Future protocols should be based on biodosimetry, as are the protocols for validating UV disinfection units for the disinfection of drinking water and wastewater. Biodosimetry means challenging the device with a target microorganism and comparing the viable counts with and without UV exposure. The SLR is determined for a range of UV exposures. In a separate experiment, the UV sensitivity of the target microorganism expressed as an inactivation rate constant or DRD D_{10} value is determined from an UV dose–response curve. Using this standard curve, the UV dose applied for each exposure in pilot or commercial unit can be determined.

Bench top scale CB units should be used to conduct biodosimetry studies. CB devices equipped with low or medium pressure lamps, pulsed lamps, and LEDs are available and shown in Fig. 4.1.

Using CB units have following essential advantages:

1. Measurable UV irradiance on the surface of product or material,
2. UV irradiance can be adjusted by changing the distance between light source and treated surface,
3. Commercial wavelength can be used,
4. Tool to conduct dose–response studies for challenge organisms on the product surface,
5. Allow conducting studies to determine the most UV-resistant organisms that can be used as a target indicator(s) in validation studies, and

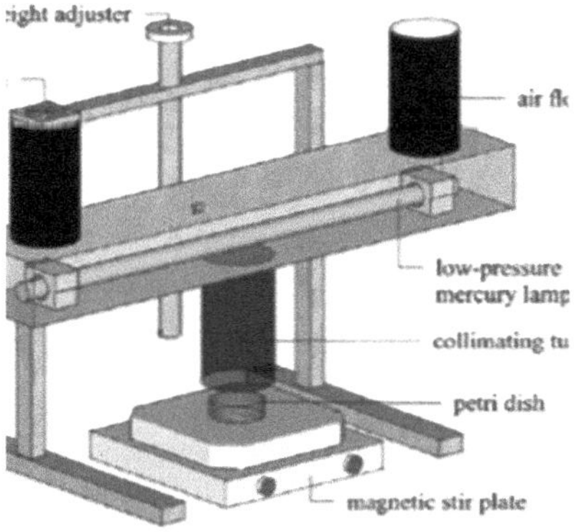

(A)

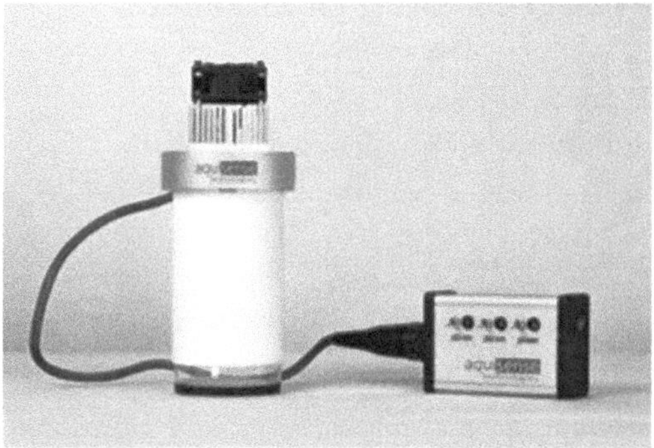

(B)

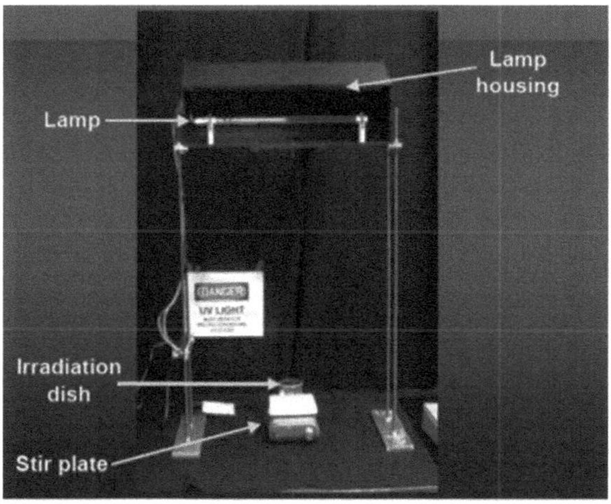

(C)

Figure 4.1 Collimated beam units equipped with various light sources: (A) continuous ultraviolet-C at 253.7 nm, (B) UV light-emitting diodes, and (C) pulsed UV lamp.

6. Allow measuring surface effects on inactivation rate and select worst case scenario of surface parameters.

4.4.4 Critical process and control parameters

During validation of light-based surface treatment process, there are a few critical parameters that will affect UV dose delivery to the product. These parameters need to be measured and recorded during validation and verified during production runs.

1. UV lamp or LEDs output irradiance; number of lamps or LED, lifetime,
2. UV intensity on the product surface. Maximum and minimum value of UV intensity have to be measured along with UV intensity distribution on the product surface or transportation means,
3. Distance between light source and product surface,
4. Exposure time,
5. Cleaning of UV lamps sleeves,
6. Temperature on the product surface, and
7. Variability of surface characteristics.

4.4.5 Commercial UV and pulsed light systems for surface treatment

There are a large number of manufacturers of UV bulbs and UV systems that are typically supplied with automated, online wipers that keep lamps free from fouling. UV lamps require replacement every year, and the quartz sleeve that houses the lamps every 2–3 years. The routine downtime per system is typically 1–2 h, and most manufacturers provide factory training for customers' technical staff if required.

The pioneer company producing PL equipment for disinfection was PurePulse Technologies Inc. (San Diego, CA), a subsidiary of Xenon Corp. Applications included water purification systems and virus inactivation systems for biopharmaceutical manufacturers. Nowadays, there are three commercial companies producing pilot scale and commercial systems based on PL: Claranor from France, SteriBeam Systems from Germany, and Xenon Corporation from United States. The detailed Information regarding devices for industrial applications can be found at the websites of these companies.

Among the commercially available system to decontaminate surfaces of foods are conveyor belt UV systems from SterilAir, Switzerland, and

Premium Light Module from Heraeus Noblelight, GmbH (Hanau, Germany). Premium UV Modules have developed by Heraeus for food industry. Disinfection modules can be installed in filling and closing machines for dairy products and beverages and used to disinfect conveyor belts, transport containers, and work surfaces and the surfaces of food. The system achieves a disinfection rate in excess of 99.9% with typical working distances of 20 mm and exposure times of 2–10 s. The packaging materials treated by the system achieve a degree of purity that corresponds to the classification as "low-germ packaging materials." Premium UV Module UV disinfection modules are ready-to-operate assembly systems consisting of one or more UV cassettes, LPA UV sources, fan housing, forced cooling with ambient air, and power supply. UV irradiance at the beginning of lifetime at a distance of 20 mm is 65 mW/cm^2 at a distance of 100 mm is 45 mW/cm^2, and at a 200 mm ~26 mW/cm^2. Premium UV Module meets IP-68 classification and thus is suitable for CIP wash-down environments.

4.4.6 Case studies of pulsed and continuous UV surface treatment

PL was tested for its utility to improve the microbial quality and safety of RTE vacuum-packaged ham and bologna slices inoculated with *L. monocytogenes*. PL treatment at 8.4 J/cm^2 reduced *L. monocytogenes* by 1.78 CFU/cm^2 in cooked ham and by 1.11 CFU/cm^2 in bologna. PL at 8.4 J/cm^2 did not affect the sensory quality of cooked ham, while treatments above 2.1 J/cm^2 negatively influenced the sensory properties of bologna. The combination of PL and vacuum packaging provided ham with an additional shelf-life extension of 30 days compared with only vacuum packaging. In addition, the effectiveness of PL was demonstrated for treatments of RTE chicken frankfurters (Sommers et al., 2009).

The packaging technologies play important role in extending the shelf-life and protecting finished food products from postcontamination during storage. Pulsed and UV light can be applied as pre- or postpackaging technology to address issues associated with cross-contamination and microbial spoilage. As a prepackaging control measure, UV and pulse light treatment of packaging in filling plant, for example, lids, cups, and sealing and packaging foils for drinks and beverages, help to eliminate microbial contamination and thus extend shelf-life of products.

Claranor (Avignon, France) offers systems using pulse light to decontaminate tubs and cups in filling lines that handle sensitive products. The

Claranor's installation consists of the electronics bay powered by the main current that generates electrical pulses and the optical cavity that focuses light toward the surface that needs to be treated. The integrated cooling system regulates the temperature of the water in the lamp circuit. According to the manufacturer, the pulse light technology guarantees a homogeneous treatment and is compatible with the high output rate of 60 strokes/min. The Claranor's systems can treat a wide range of types and sizes of cups for preformed and formed fill seals in bottles up to 15 cm depth. It is perfectly adapted for thermally treated products stored in cold chain for cups' sterilization of fruit-based desserts, cups' disinfection of dairy products, juices, and ready meals filled at cold temperature. The application of PL for packaging sterilization for thermally treated products packaged in aseptic conditions, such as soups, sauces, creams, fruit preparations, and stewed fruits, has been developed.

Among known existing applications of UV light in meat processing are UV treatments of nonfood and food contact surfaces and recently product surfaces. UV light was effective against *E. coli* and *Salmonella* on pork skin, raw meats, and poultry surfaces. The precautions should be taken to avoid undesirable brown discoloration of red meat since UV encourages metmyoglobin formation. The rate of product discoloration depends on the intensity of light and wavelength distribution. Packaged product can also be exposed to UV so to avoid of risk of recontamination. The correct choice of polychromatic pulsed or monochromatic continuous source in combination with the UV filter and light permeability of the packaging material is critical when such a process is developed for red meat products.

According to US Department of Agriculture (USDA), contamination of foods with *L. monocytogenes* is likely to occur in all RTE meat and poultry products that are exposed to postlethality processing environment. These postlethality processing environments are the areas into which products are routed after complete treatments and may include slicing, peeling, dicing, rebagging, and brining. As an example of UV technology postlethality application, packaged RTE product can be exposed to UV so to avoid of risk of cross or recontamination. Knowledge of light permeability of the packaging material is critical when such a process is developed.

USDA scientists (Sommers et al., 2009) have demonstrated that UV light can be used to reduce levels of *Listeria* contamination on the surfaces of RTE products. UVC irradiation of frankfurters that were surface-inoculated with *L. monocytogenes* resulted in a 1.31-, 1.49-, and 1.93-log reduction at doses of 1, 2, and 4 J/cm^2, respectively. UVC treatment had no effect on frankfurter color or texture at UVC doses up to 4 J/cm^2. It was concluded that because the numbers of *L. monocytogenes* associated with contaminations of RTE meats are typically very low, the use of UVC in combination with potassium lactate and sodium diacetate has the potential to reduce the number of frankfurter recalls and food-borne illness outbreaks (Sommers et al., 2010).

In recent studies, UV light was reported to be effective against nonspore-forming pathogenic microflora on raw meats and poultry surfaces. UVC at 0.5 J/cm^2 reduced the initial populations of *Campylobacter jejuni*, *L. monocytogenes*, and *Salmonella typhimurium* by 1.3–1.2 log CFU/g on chicken breasts (Chun et al., 2010). Similarly, UV treatment of raw chicken fillet at 0.2 J/cm^2 reduced *C. jejuni*, *E. coli*, serovar *Enteritidis*, total viable counts, and Enterobacteriaceae by 0.76, 0.98, 1.34, 1.76, and 1.29 log CFU/g, respectively. Following UV treatment of packaging and surface materials, higher reductions of up to 3.97, 4.50, and 4.20 log CFU/cm^2 were obtained for *C. jejuni*, *E. coli*, and serovar *Enteritidis*, respectively. These studies indicate that UV may be beneficial in working areas of meat and poultry processing plants to reduce the level of aerobic and nonspore-forming pathogens and may be applied to decontaminate cut-up products moving on conveyors, associated packaging, and surface materials.

Review of the reported studies and commercial implementation trials confirmed that light technology has a great potential to aid in food industry efforts to produce safe products with ESL and provide additional barrier of foods security on consumer shelves. Commercial UVC systems can be readily integrated in food production facilities at various points of production and assist in addressing safety and extended storage objectives.

4.5 Validation of UV light preservation processes for liquid products and beverages

Applications of UVC light is emerging in food and beverage processing as a nonthermal, nonchemical, and nonionizing organic method of preservation and safety that offers a less expensive, energy-saving alternative solution to heat for the treatment of liquid ingredients, juice products, other beverages, and milk and dairy products. Compared to another nonthermal process, such as high hydrostatic pressure (HHP), UVC has an advantage of continuous mode treatment and offers packaging flexibility. A large number of reported data have shown the effectiveness of UVC light at 253.7 nm against food-borne pathogens in liquids, natural microbiota, molds, and yeasts. Also, due to its nonthermal nature, UVC processing resulted in better product quality, higher nutritional content, and fresh-like sensory attributes of products compared to the thermal processing. However, despite these obvious benefits, existing regulatory approvals of UV treatments of juices in the United States and Canada and milk in EU, successful experience of application of UVC for municipal and wastewater treatment, and commercialization of UVC applications for opaque liquids and beverages have been slow. Among a few causes of the current state are the lack of understanding of UV light transmission through opaque medium or liquids with low UV transmittance (UVT) and differences in the approaches of evaluation of UV energy, fluence, and dose in this instance in comparison with well-established approaches in water treatment with high UVT. Consequently, there are no established practices on the validation of microbial performance, fluence, or UV dose delivered in continuous systems for the treatment of low-UVT liquids. Different theoretical terms often used in the studies to report lethal action and efficiency of UV light, thus creating a situation where data comparison becomes problematic.

This section will discuss the knowledge gaps and differences in UV light penetration when opaque liquids or liquids with low UVT are treated compared with water. Also, the essential differences in water and beverage product characteristics, UV energy, fluence and dose determination, and process requirements will be presented and discussed to be considered and applied for successful UVC process development and validation. The essential steps in product and UVC process development will be summarized that need to be implemented to accelerate UV technology commercialization for low-UVT liquids. This will include major requirements for UVC preservation of juices and other beverages, principles of establishment of equivalent preservation UVC dose with the consideration of quality and sensory effects, challenges of process validation and controls and effective designs of UVC systems.

4.5.1 Requirements for UVC preservation of beverages

Developing a new UVC process requires initial establishing a proper design and operational doses followed by a validation step to ensure that

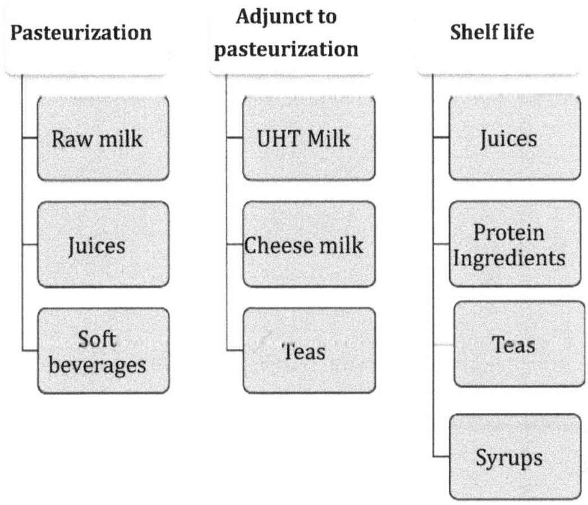

Figure 4.2 Examples of applications of ultraviolet-C technology for the preservation of opaque beverages: pasteurization, adjunct to pasteurization, and shelf-life extension.

UVC dose is reliably delivered to the product and evaluation of the microbial performance of a system. Unlike in water treatment, the fluence and UVC dose for beverages has not been established by the regulators, and the process has to be developed based on achieving product-specific process requirements in terms of inactivation organisms of concern or so-called intended technical effects. The intended technical effect for liquid foods and beverages can require addressing food safety issues often through a pasteurization step and achieving shelf-life extension at refrigerated or ambient storage conditions. This can involve achieving a 5-log target pathogen reduction in juice products to comply with hazard analytical critical control points (HACCP) (US FDA, 2001), shelf-life extension of fresh juices, or adjunct treatment for pasteurized milk, teas, or liquid sugars by eliminating heat resistant spores or spoilage organisms as shown in Fig. 4.2.

Since preservation and storage requirements are very much defined by the product pH, the examples of establishment of preservation specification for three pH categories of high acid ($pH < 3.5$), acid ($3.5 < pH < 4.6$), and low acid (pH.4.6) of juices and soft beverages are shown in Table 4.3.

In the case of using UVC light for milk preservation, the process requirements will be defined by the specific operation in milk processing where UVC can be applied, such as thermization, pasteurization, ESL

Table 4.3 Product, process, and storage conditions for the establishment of preservation specifications of juices and beverages.

Product pH	High acid pH < 3.5	Acid 3.5 < pH < 4.6	Low acid pH > 4.6
Additional hurdles	No Carbon dioxide	Refrigeration Carbon dioxide	Antimicrobials Water activity
Pathogenic organisms	*Escherichia coli ListeriaSalmonella*	*Escherichia coli ListeriaSalmonella*	*Nonproteolytic Clostridium botulinum*
Specific log reduction	5 log	5 log	5 log
Indigenous microbiota	Total aerobic/anaerobic plate count	Total aerobic/anaerobic plate count	Total aerobic/anaerobic plate count
End point concentration	<100 CFU/mL	<100 CFU/mL	< 100 CFU/mL
Spoilage organisms	Aciduric bacteria <10 CFU/mL Coliforms <10 CFU/mL Molds, yeasts <10 CFU/mL Lactic acid bacteria <10 CFU/mL	Aciduric bacteria <10 CFU/mL Coliforms <10 CFU/mL Molds, yeasts <10 CFU/mL Lactic acid bacteria <10 CFU/mL	
Storage	Ambient Refrigerated	Refrigerated conditions	
Packaging	Hermetically sealed containers Minimum head space		

treatment, or sterilization step for chilled or ambient distribution. Also, the UV process should be optimized to achieve the best quality, nutritional, and sensory attributes. After the UVC operational dose is established, its effects on product enzymes and stability of the suspension should be taken into consideration for improved product quality during storage.

4.5.2 Establishment of preservation reduction equivalent UVC dose

A general approach for the establishment of the UVC preservation process for liquids includes the following three steps.

1. Identification of the organisms of concern (pathogenic and spoilage) and their UVC resistance

   The choice of the organism of concern in juice and other beverage products depends on product pH and will differ for high-acid, acid, and low-acid beverages as shown in Table 4.3. Knowledge of the food formulation and history of the food outbreaks and/or evidence of potential growth is essential when selecting the appropriate challenge pathogens. Multiple specific strains of the target pathogens should be included in the challenge study and their UVC resistance has to be measured to select the most UVC resistant strains. It is typical to challenge a food formulation with a "cocktail" or mixture of multiple strains to account for potential strain variation. The ideal organisms for challenge testing are those that have been previously isolated from similar formulations and pathogens from known food-borne outbreaks should be included to ensure that the formulation is robust enough to inhibit the growth of these organisms. It is not unusual to have a cocktail of five or more strains of each target pathogen in a challenge study. A bench top CB setup can be used to measure UVC resistance in the tested medium. However, the essential modifications to the standard water protocol have to be made due to high UV absorbance of beverages and to achieve uniform UVC dose delivery in the sample. Currently, in the absence of established research methods, standard approaches, biosafety level 2 and 3 labs, as well as resources to conduct microbial UV inactivation studies in opaque beverages, a large variance in UVC resistance of food-borne pathogens can be observed among reported data. However, in the proper setup and correctly designed biodosimetry experiments, a first-order linear model adequately describes the destruction of organisms for many beverages under the action of UVC or inactivation rate constants should be determined using adequate nonlinear models.

2. Identification and selection of the appropriate reduction target—safety and shelf-life

As indicated above, for high-acid juices, the 5-log reduction of pathogenic *E. coli*, *Salmonella*, and *Listeria* has to be demonstrated to comply with US FDA HACCP rules. For low-acid products, the process has to be capable of reducing 5 logs of nonproteolytic *Clostridium botulinum*. With respect to reduction requirements of spoilage and natural microbiota, the "normal" initial levels seen in beverages processing operations have to be evaluated and the recommended "end point" approach should be used to determine the margins of the process (Table 4.3).

An appropriately designed microbiological challenge test will validate whether a specific process is in compliance with the predetermined performance standard, for example, a 5-log reduction of *E. coli* O157:H7 or other pathogens for fresh juices. Table 4.4 summarizes the examples of pasteurization conditions for juices and other products stating product pH,

Table 4.4 Examples of pasteurization process for different pH groups: juices, milk, and liquid eggs.

Examples of products	pH	Pathogen of concern	Specific microbial reduction (logs)	Enzymes' destruction
Apple cider	< 3.5	*Escherichia coli* O157:H7	5 log	
Orange juice	<3.5	*Salmonella*, *Escherichia coli* O157:H7	5 log	Pectin-methylesterase
Carrot juice	> 4.6	Nonproteolytic *Clostridium botulinum*	5 log	
Milk and milk products	6.5–7	*Mycobacterium tuberculosis*; *Coxiella burnetii*	5 log	Negative for alkaline phosphatase
Egg products	> 7	*Salmonella enteritidis*; *Salmonella typhimurium*	7 log	

target pathogen of concern, and required SLR based on regulatory requirements and juice HACCP. The choice of the organism of concern in juice depends on product pH. For high-acid juices, the 5-log reduction of pathogenic *E. coli*, *Salmonella*, and *Listeria* has to be demonstrated. For low-acid products, the process has to be capable to reduce 5 logs of non-proteolytic *C. botulinum*.

The possibility of growth of selected organisms in target products should also be explored. Records of normal microbial counts in industrial food products are known for their irregular, fluctuating character that is determined primarily by variations in the initial load and numerous random factors, which tend to promote or inhibit the microbial growth. Since our understanding of the complex and dynamic microbial ecosystems in food production environments is still relatively superficial, food preservation and safety assurance systems must remain crude and be guided by general principles that apply to both spoilage and pathogenic species.

3. Development of a conservative estimation of UVC dose to deliver the targeted microbial reduction

   Preservation reduction equivalent UVC dose (H_{germ}) has to be determined and applied to meet a required SLR in numbers of target pathogenic organisms or "end point" concentration of spoilage organisms using Eqs. (4.4) and (4.5)

$$H_{germ1} = D_{10uv} \times \text{SLR} \tag{4.4}$$

$$H_{germ2} = D_{10uv}(\log N_0 - \log N_F) \tag{4.5}$$

According to Eqs. (4.3) and (4.4), H_{germ1} is the amount of UVC energy in mJ/cm^2 that has to be delivered to the target pathogenic organism(s) in the product to achieve targeted SLR.

D_{10uv} is a UVC dose to reduce 90% of relevant pathogenic population with the highest UVC resistance. The germicidal dose H_{germ2} is the UVC dose needed to achieve the end point concentration of spoilage organisms and is a function of the initial load (N_0), the end point of the process (N_F), and the resistance of target spoilage bacteria under defined conditions (D_{10uv}). It was reported that typically the UVC resistance of spoilage yeast and mold spores is much higher than pathogenic bacteria. Also, it is recommended that the highest value of H_{germ1} or H_{germ2} be used as the UVC preservation RED.

4. Determination of the critical product parameters to control the delivery of the required process

The actual preservation UVC dose can vary for different categories of juices, beverages, dairy products, and liquid ingredients and very much depends on product composition, absorption and scattering properties, rheological characteristics, initial microbial load, and reduction requirements. Within a food product, several factors can influence the delivery of the UVC dose, such as pH, soluble and suspended solids, and product composition. The presence of particles could provide a means of shielding microorganisms from UV light photons.

4.5.2.1 Physico-chemical and optical properties of opaque liquids

A few factors found to consistently affect the efficacy of UV inactivation in opaque beverages were product pH, Brix, absorbance and scattering or product UVT, viscosity, and temperature.

4.5.2.2 pH, Brix, UV transmittance, turbidity, and viscosity

Product pH, Brix, UVT, turbidity, and viscosity are critical product parameters that can affect UV photons' delivery and subsequently lethal action of UVC light.

pH or measure of the acidity or basicity of the products, dissolved solids (Brix), and water activity are considered as hurdles that can significantly modify the efficacy of UVC inactivation.

Fluid chemical composition and concentration of dissolved and suspended solids determine whether the product is transparent, opaque, or semitransparent.

Optical properties, including absorption and scattering, along with the design of the UV systems, are the major factors affecting UV microbial inactivation in fluids with high absorptive properties or low UVT. The UV absorbance (A) strongly influences the UVC dose delivery to microorganisms. The UV absorbance is a function of the concentration of an absorber of electromagnetic radiation (c), extinction coefficient (ε), or molar absorptivity of the absorbing species (L/mol/cm), which is a measure of the amount of light absorbed per unit concentration absorbance or optical density, and path length of light (d).

For low-UVT liquids, turbidity is another critical product parameter that should be controlled. Particle content, association, and clumping of microorganisms can also affect UV performance. Particles can absorb and scatter UV light to varying degrees based on their size and composition. In general, particles affect the UV process in two ways:

1. Particles can decrease the UVT of liquid and thereby affect the UV dose delivery.
2. Microorganisms can associate with particles and be shielded from UV light, thereby changing the characteristics of the UV dose–response curve that is obtained using CB studies.

Changes and variations in UV absorbance of food products should be considered and a simple absorbance reading may prove inadequate to predict the microbial destruction within most beverage and liquid systems.

Other physical properties of fluids, such as viscosity and density, influence the effectiveness of momentum transfer and define flow pattern in the UV system. In addition, viscosity of product impacts the hydrodynamic behavior of the liquid in the UV system and, consequently, the delivery of UV photons. The effect of product's temperature on viscosity should also be considered.

Characterization of physical, optical, and biochemical properties of liquids is highly desirable for proper process establishment. Sufficient UV absorbance data of liquids should be properly collected, characterized, and reported. The absorbance measurement techniques should account for the presence of the suspended particles and soluble solids and their effect on the estimation of the UVC dose. In addition, the effect of some essential compounds, such as vitamin C, on the absorption effects of the product needs to be taken into account during UV treatment. In the determination of the UV absorbance design criteria, the full range of UV absorbance at the UV implementation point where the system will be installed should be determined. This enables selection of an appropriate UVT range to ensure compliance at all times.

4.5.3 Principles of validation of UVC preservation process for liquids and beverages

The primary objective of validation is to demonstrate UVC process consistency to deliver the required performance objective (safety objective) over time across R&D scales and manufacturing environment with various raw materials and operating ranges.

Similar to established practices in UV water treatment, dose delivery by the system can be assessed using biodosimetry. With biodosimetry, inactivation of challenge microorganisms is measured and related to a 90% reduction dose or RED using the known UVC dose–response. Since there are a few types of UV doses/fluencies that typically have to be determined in biodosimetry studies, it is important to define terminology for the UV treatment of opaque and low-UVT liquids and beverages.

4.5.3.1 UV energy, fluence, and germicidal dose in water and opaque liquids

In validation studies, the exposed and absorbed dose of UVC process will define microbial efficacy and impact on essential products attributes and quality. Also, to properly evaluate the efficacy of any UV process, it is necessary to classify UV treatment in terms of the energy or output power data information (from manufacturer or measured) and according to the UV source, the system design, and product absorptive characteristics. The difference in determination of UV light parameters in a batch and continuous flow system should be considered.

The following terminologies for the characterization of UV light lethal action are often used: total UV energy, applied UVC energy, incident fluence, average absorbed fluence, and delivered or germicidal fluence (dose). The essential differences between these system and process characteristics shall be clearly understood and used appropriately in UV process calculations. The schematic illustration of the UV energy–fluence–dose concept in water and liquid with low UVT in batch setup is presented in Fig. 4.3.

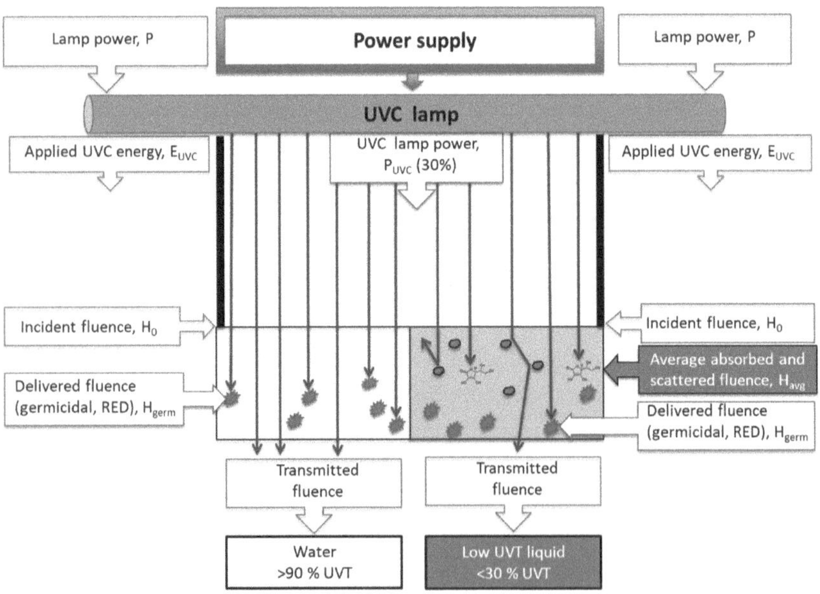

Figure 4.3 Graphical illustration of the concept of ultraviolet (UV) lamp power to UV dose in water and low-UV transmittance liquid in batch UV setup.

4.5.3.2 UV Batch system

In typical batch UV system setup (Fig. 4.3), UV light sources are installed at the distance from a treated sample. The total applied energy (E) (J/cm^2) can be determined using Eq.(4.6), knowing the output power (W) of the UV source:

$$E = \frac{P}{A_l} \times t \tag{4.6}$$

where P is the total output power of the UV source (W), A_l is the surface area of the UV lamp quartz sleeve (cm^2), and t is the time (exposure time) (s).

The output power in the UVC range from 200 to 400 nm, (P_{UVC}), is typically around 30% or 10% of the total output wattage for LPM and medium-pressure mercury (MPM) lamps, respectively, meaning that the applied energy in UVC range (E_{UVC}) can be calculated using Eq. (4.7)

$$E_{UVC} = \frac{P_{UVC}}{A_l} \times t \tag{4.7}$$

P_{UVC} is the output power of the UVC source in UVC range. Typically the UVC efficacy of LPM and MPM lamps is 30% or !0% of the total wattage of these lamps respectively.

In batch UV systems, the *incident* fluence (H_0), in J/cm^2, is often reported and can be defined as exposed dose and is the energy that is actually incident at the surface of the product. H_0 is calculated based on the incident UV irradiance or fluence rate in W/cm^2, on the surface of the product using Eq. (4.7). Fluence rate is defined as the total radiant power incident from all directions onto an infinitely small sphere of cross-sectional area dA, divided by dA (Bolton and Linden, 2003) and incident UV fluence is the fluence rate multiplied by the exposure time. However, due to the fact that in CB systems only beams that are perpendicular to the sample surface can be measured, the incident intensity (I_0) is used in the calculations (Eq. 4.8).

$$H_0 = I_0 \times t \tag{4.8}$$

I_0 is the INCIDENT UV intensity, radiometer reading at the center of the product and at a vertical position so that the calibration plane of the detector head is at the same level as the surface of the fluid (W/cm^2). I_0 can be provided by the manufacturer, but its value should also be measured with a radiometer at a controlled distance from the UV source and the measurement has to be checked periodically to take into account the decrease in lamp power during its lifetime.

Incident fluence (H_0) depends on the configuration of the system, but it is independent of the material being irradiated. Total exposure time is considered in the calculations of batch systems, either agitated or in a thin layer

The volume average absorbed fluence or dose (H_{avg}) in the liquid can be determined knowing absorbed UV irradiance or fluence rate ($I_{r\ avg}$) using general Eq. (4.9).

$$H_{avg} = I_{avg} \cdot t \tag{4.9}$$

Transmitted UV fluence is a remaining fluence (Eq. 10) that was not absorbed by the liquid medium and microorganisms and depends on absorbing and scattering properties of the product and Petri dish material.

$$H_{trans} = H_0 - H_{avg} - H_{germ} \tag{4.10}$$

Harm (1980) considered several types of liquids as typical examples. They can be characterized as homogeneous solutions (transparent, semi-transparent, and opaque) and particulate samples or suspensions. If sample is a homogeneous liquid, light absorption in homogeneous solutions follows relatively simple laws. However, very often the suspensions of whole cells are irradiated which can be considered as dense packages of absorbing molecules separated from each other by the suspension liquid. The suspension liquid itself may not absorb UV radiation as is the case if the liquid is water, an inorganic buffer solution or a glucose—salts medium. In homogenous solutions, if a uniform layer is exposed to a parallel beam of monochromatic UV radiation at a fluence rate I_0, the fluence rate of the unabsorbed transmitted fluence rate emerging after passage through the layer is expressed by Eq. (4.11):

$$I_{trans} = I_0 10^{-A} \text{ or } I_{trans} = I_0 e^{-Aln10} \tag{4.11}$$

where A is the absorbance of material.

If the broadside area of the solution, S, is exposed to the incident fluence H_0, which is the product of fluence rate and the time, I_0t, the equation to calculate absorbed fluence can be obtained from Eq. (4.12):

$$H_{abs} = (I_0 - I_{trans})tS = I_0t\left(1 - 10^{-A}\right)S = I_0t\left(1 - e^{-Aln10}\right)S \tag{4.12}$$

In batch-CB UV systems for water ($A << 1$, $UVT > 90\%$), Bolton and Linden (2003) recommended using Eq. (4.13) to determine the average absorbed portion of fluence rate in water

$$I_{avg} = I_0 \times RF \times PF \times WF \times DF \tag{4.13}$$

where WF is the water factor or liquid factor is calculated according to Eq. (4.14)

$$\mathrm{WF} = \frac{1 - 10^{-a_\lambda l}}{a_\lambda l \quad \ln(10)} \tag{4.14}$$

α_λ is the decadic absorption coefficient of liquid or absorbance of this liquid for a 1-cm path length at a λ (nm) wavelength (1/cm). DF is the divergence factor; for path lengths less than 5 cm, the DF is given by Eq. (4.15)

$$\mathrm{DF} = \frac{L}{(L + l)} \tag{4.15}$$

R is the reflection factor ($RF = 1 - R$). For air and water: $R = 0.025$ and $\mathrm{RF} = 1 - 0.025 = 0.975$. PF is the Petri factor, measured according to Bolton and Linden (2003). l is the vertical path length of the sample in the container (such as a Petri dish) (cm). L is the distance from the UV lamp to the surface of the sample (cm).

In the nearly opaque case, as it follows from Eq. (4.6), virtually all incident UV fluence entering the solution is absorbed ($A > 2$, UVT < 1%) meaning that transmitted fluence is close to 0 and can be calculated using Eq. (4.16).

$$H_{\mathrm{abs}} = H_0 S \tag{4.16}$$

The intermediate or semitransparent case ($1 < A < 2$, 10% < UVT < 90%) is represented by a solution that absorbs a measurable fraction of the incident photons but lets the remained light pass. If the solution is well stirred so that all molecules are statistically equally exposed to the radiation, such a case can be treated for the purpose of the calculations as a near-transparent solution receiving a lower fluence (Eq. 4.12).

The *delivered* or germicidal UV fluence or dose (H_{germ}), J/cm^2, is the remaining available energy in the product (that has not been absorbed by the other constituents of the fluid) that is actually delivered to the microorganisms. The delivered or germicidal fluence is typically estimated by using biodosimetry approaches that will be discussed later.

The relationship for the above energies in low-UVT liquids can be expressed as: UVC incident dose = UVC absorbed dose (medium) + UVC germicidal dose (delivered) + UVC transmitted dose + UVC reflected dose (not shown in Fig. 4.3 but accounted in calculations).

4.5.3.3 UV Continuous systems

In continuous UV systems, the total applied UV and UVC energy (E_{UV} and E_{UVC}) can be expressed in J/L (Eqs. 4.16 and 4.17). It is calculated based on the total output power of the UV source (P_{UV}), in W. Such data are given by the manufacturer or by the equipment energy consumption information. The time used in the calculation is the fluid volume average residence time in the UV system. E_{UV} is independent of the material being irradiated and of the configuration of the system. It is important for researchers reporting applied UV energy in continuous systems to supply all the necessary system parameters required for energy calculation in both J/L and J/cm^2. Interconversion between these units will allow for proper comparison of UV dose values among researchers and commercial applications. The interconversion has to be done for each system.

$$E = \frac{P}{\dot{V}} = \frac{L_N \times P}{\dot{V}} \tag{4.17}$$

$$E_{UVC} = \frac{P_{UVCN}}{\dot{V}} = \frac{L_N \times P_{UVC}}{\dot{V}} \tag{4.18}$$

L_N is the number of the UV sources with output power P (W), and V is the volumetric flow rate (m^3/s).

For comparison purposes, the applied UV energy in continuous UV systems can be expressed in J/L; the applied UV dose in batch systems was often expressed in J/cm^2.

The volume average absorbed UV fluence (H_r), in J/cm^2, takes into account geometrical parameters of the system and optical properties of the liquid product. H_r is the radiant UV energy that is actually absorbed by the medium flowing in the chamber and its constituents (homogeneous fluid) and is available for driving a photo-reaction. H_r is calculated based on the absorbed UV irradiance or fluence rate (I_r), in W/cm^2, by the product on exposure time t (Eq. 4.19). I_r in continuous system can be calculated using Eq. (4.20) that accounts for the geometry of the system and optical properties of the medium. The time used in the calculations is the UV exposure time or average period of time in which each volume of the liquid is in close contact with the surface of the lamp sleeve. Often exposure time can differ from actual residence time in the system chamber.

$$H_r = I_r \times t \tag{4.19}$$

H_r depends on the UV light-absorbing characteristics (absorption coefficient) of the material being irradiated and the configuration of the system. Absorbed UV irradiance or fluence rate (I_r) in a continuous flow UV system can be calculated using Eq. (4.20)

$$I_r = I_0 \times \frac{r_0}{r} \times \exp[-\alpha_\lambda \times (r - r_0)] \tag{4.20}$$

α_λ is the decadic absorption coefficient or absorbance for a 1-cm path length at a λ (nm) wavelength (1/cm). r is the radius of the chamber (cm). r_0 is the radius of the UV lamp quartz sleeve (cm). I_0 is the incident fluence on the surface of the lamps sleeves, mW/cm^2.

Also, absorbed fluence can be evaluated using UV sensors built in the UV systems and measured the transmitted UV fluence in the products. Due to complexity of geometrical configuration, flow patterns and trajectory of organisms within the UV chamber, a computational fluid dynamics (CFD) modeling can be used to evaluate UV fluence in various liquids along with UV inactivation efficiencies. By using the CFD modeling, the irradiance modeling can be coupled with fluid dynamics and simulated the trajectories of microorganisms through the UV system. This will allow determination of real exposure time of organisms to UV light.

Delivered or germicidal fluence or dose (H_{germ}) in continuous flow UV system is determined using biodosimetry approach (Eq. 4.21) based on the log count reduction of organisms or (LCR = log N_0−log N_F) measured in the tested liquid in the system and fluence for 1 logarithmic reduction (D_{10uv}) determined from biodosimetry experiments in CB tests using the same liquid medium and organisms.

$$H_{germ} = D_{10uv}(\log N_0 - \log N_F) = D_{10uv}\mathrm{LCR} \tag{4.21}$$

N_0 and N_F are the initial and final concentrations, respectively, of microorganisms in tested liquids (CFU/g).

Determination of UV energy, fluence, and doses was not clearly reported in many studies. The above equations are to be considered as suggested approaches, once there are other ways to calculate and measure the absorbed UV irradiance and germicidal dose.

4.5.3.4 Terminology and evaluation of UV dose in water treatment practices

For continuous-flow reactors, the terms "reduction equivalent dose," "design UV dose," and "operational UV dose" are used in drinking water treatment practices and defined as follows: RED is the dose that is assigned to the UV test reactor based on reactor validation testing. The RED is equivalent to that measured with the CB apparatus for the same degree of inactivation of the target microorganism. Establishing the RED in continuous flow system involves determining the UV inactivation of a selected microorganism under controlled batch conditions using CB system. In the case when flow processing conditions in the commercial unit are known, bench-top flow-through unit can be built to determine RED in similar flow mixing conditions (equal Reynolds numbers), product UVT, and lamp source.

Design UV dose. The RED is required for a specific log inactivation of the target microorganism. The design UV dose is used for sizing UV disinfection systems. The design UV dose will depend on the target microorganism and the quality of the water-supply source prior to UV disinfection. The target microorganism and the required inactivation levels must be established by the regulatory agency.

The design RED in mJ/cm^2 in the flow-through system can be determined for each processing condition using the measured log inactivation and the UV dose–response curve developed through CB testing (Fig. 4.4—RED linear curve, step 1a).

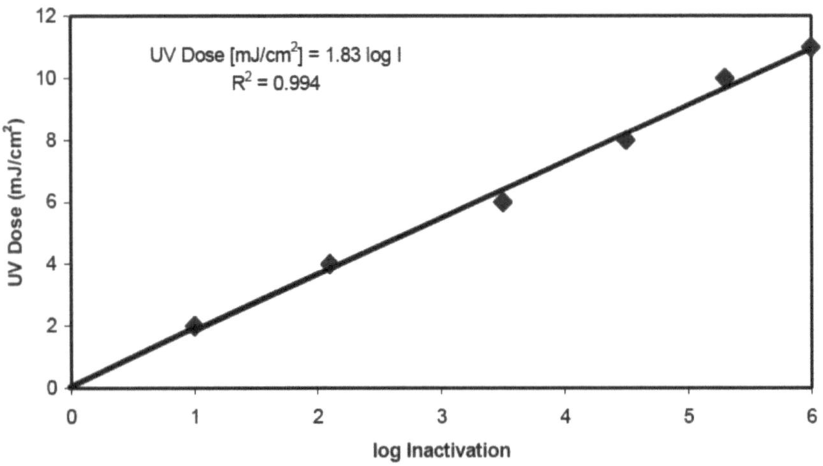

Figure 4.4 Linear ultraviolet (UV) dose–response curve from collimated beam tests—step 1a. (Note: linear curves are typical for high UV transmittance and nonlinear for low UV).

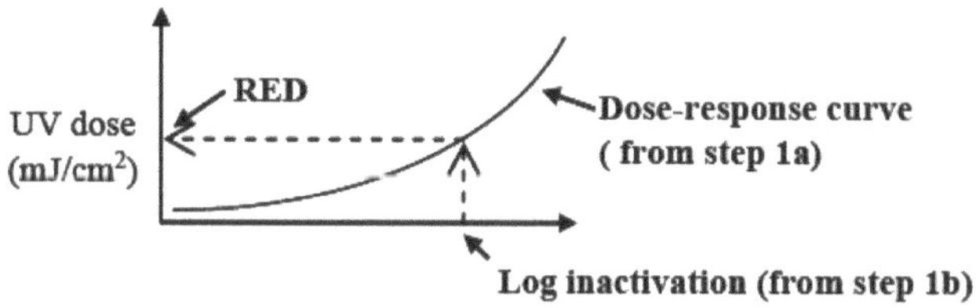

Figure 4.5 Determination of reduction equivalent dose (RED) in continuous flow ultraviolet (UV) unit: input log inactivation from step 1b into the UV dose–response curve from step 1a (nonlinear curve).

Fig. 4.5 demonstrates step 2 in the biodosimetry tests in flow-through system using RED nonlinear curve obtained in CB tests (step 1a).

Operational UV dose. The UV dose that is established based on the results of the equipment validation testing. The operational UV dose, a useful index of reactor behavior, can be used to make most efficient use of the UV disinfection system while maintaining the design UV dose. The UV intensity measurements or UV sensor reading are needed to establish the operational UV dose.

The terminology established for UV water treatment industry shall be used in UV treatment of liquids foods and beverages practices.

4.5.3.5 Biodosimetry studies and challenges for low-UV transmittance liquids

As indicated earlier, biodosimetry studies are used in UV industry to measure the delivered germicidal fluence (H_{germ}) that is also termed "reduction equivalent dose (RED)."

In biodosimetry tests, conventional UV inactivation kinetics is obtained by irradiating suspensions of cells in water, buffer, or tested liquid in well-stirred Petri dish or thin microcuvettes. The bench-scale CB device (Qualls et al., 1983, Kuo et al., 2003) has evolved as a standard method in UV water treatment. Laboratory dose–response data from CB tests are commonly used as a basis for determining the necessary delivered UV dose or RED for full-scale UV systems. The typical UV fluence–response function plotted as log $I = \log N_0/N$ versus UV dose is shown in Figs. 4.4 and 4.5.

The UV fluence–response of microbes is determined by measuring the concentration of microbes capable of replication after exposure to a measured average absorbed UV dose in CB unit.

After conducting microbial challenge test in continuous UV system and measuring LCR (N_0 and N) or log(I) in tested liquid, the RED of UV system can be calculated using Eqs. (4.22) and (4.23):

$$UV_{dose} = A \log I \quad (4.22)$$

$$\log(I) = \log(N_0/N) \quad (4.23)$$

where A is the linear slope as shown in Fig. 4.4.

Individual UV fluence–response curve should be developed for each UVT condition tested (typically high UVT and low UVT). Environmental Protection Agency (EPA) recommends using regression analysis to develop each UV dose–response curve. For challenge microorganisms exhibiting first-order kinetics, a linear equation should be used.

Also, using linear inactivation curves, the 1-log reduction or decimal UV dose (D_{10uv}) that is characteristic measure of UV resistance can be determined for a specific challenge organism. Often the UV fluence–response curves in low-UVT liquids generated using CB unit and liquid stirring in Petri dishes are not linear due to nonuniformity of treatment. In this case, thin microcuvettes with approximate thickness equal to the penetration depth of UV light in the tested liquid can be recommended for use.

The measured RED will have a value lying between the minimum and average in the dose distribution. The inactivation of a pathogen cannot be precisely determined using bioassay data unless the reactor's dose distribution is known or the challenge microorganism has the same UV dose–response curve as the pathogen. The range of inactivation is bounded by the inactivation expected with an ideal and worst case reactor.

With an ideal reactor delivering a single dose, the bioassay is a measure of the dose delivered to all microorganisms. The log reduction of any given microorganism can be calculated using Eq. (4.24):

$$\log(S_p) = \frac{\log(S_c) \times D_c}{D_p} = \frac{\text{RED}}{D_p} \quad (4.24)$$

where $\log(S_p)$ is the log reduction of the pathogen. $\log(S_c)$ is the log reduction observed with the challenge microorganism. D_p is the UV sensitivity of the pathogen expressed as the dose required per log reduction or DRD. D_c is the UV sensitivity of the challenge microorganisms expressed as the dose required per log reduction or DRD. RED is the dose equivalent of the challenge microorganism calculated.

Eq. (4.22) can be used to calculate a specific log reduction in terms of spoilage microorganisms such as yeasts and molds that can be achieved by the designed process. Ideally, the UV reactor's performance should be challenged using a surrogate microorganism whose UV sensitivity matches that of the target pathogen as defined by regulations.

In the current situation, the number of validation studies of UVC processing of food liquid products and beverages is very limited and no guidelines or standard protocols have been developed so far to assist researchers and food processors. Also, it is important to recognize that each preservation process is unique and suggests that validation studies will be process and product-specific. It is the responsibility of the food and beverage processors and process and food safety authority to demonstrate the ability of their process to achieve the SLR of the pertinent microorganisms in the product. However, the developing standard validation guidelines should lead to significant commercial benefits for the manufacturers and UVC technology providers.

4.5.3.6 Surrogate or indicator organisms

Unless working in a controlled environment (e.g., a biosafety level 2/3 pilot scale containment facility), pathogens should not be used in an open processing environment with a potential risk for the contamination of equipment, packaging, and facilities. In these situations, nonpathogenic "surrogate" organisms or challenge microorganisms are selected for use in lieu of the pertinent pathogen. According to the US FDA (Anonymous, 2000), the surrogate microorganism is defined as "nonpathogenic species and strain responding to a particular treatment in a manner equivalent to a pathogenic species and strain." Obviously, such challenge strains must have resistance traits that have been predetermined in controlled studies to match as closely as possible those of the pertinent pathogen. If a surrogate strain is to be used in a microbiological validation study, preliminary work should be done to well characterize the strain before its use in the validation study.

E. coli O157:H7 has been known as a cause of serious illness and mortality in outbreaks when unpasteurized apple cider or apple juice was consumed. Shiga toxin-producing (stec) serogroups, such as O26:H11, O103: H2, O145:NM, and O111:NM, also were recently recognized as emerging human pathogens and were associated with several juice outbreaks. Non-O157 STEC serogroups cause indistinguishable from O157:H7-induced disease and therefore research was mainly focused on *E. coli*

O157:H7. However, since UV resistance may vary among the microorganisms in the food matrixes, the most UV-resistant pathogenic strain of O157:H7 group has to be determined and used for the establishment of UV process.

On the base of the suggested selection criteria for surrogates, Orlowska et al. (2014) developed a systematic procedure for the selection and justification of the proper surrogate organism that can be used to validate novel technologies based on UV light. This procedure included seven essential steps:

1. Selection of several pathogenic strains of health concern identified for a given food product or beverage.
2. Selection of several nonpathogenic strains, which are
   a. nonpathogenic to testing personnel if handled properly,
   b. phylogenetically close to the pathogen of concern, and
   c. capable of coexisting with the pathogen in the food matrix.
3. Comparative survival studies of pathogenic and nonpathogenic strains in the given food matrix (pH, water activity, composition). The nonpathogenic strains characterized by equivalent or higher survival than pathogens shall be identified and used in further testing.
4. Comparative UV screening studies at UV sources wavelengths using both pathogenic and nonpathogenic strains selected in step 3. In step 4, a few nonpathogenic strains characterized by equivalent or higher UV resistance at wavelength of the source than pathogens in a given food product are determined.
5. Challenge UV inactivation studies of food product inoculated either with single strain or cocktail, that is, mixture of pathogenic and nonpathogenic strains, which were selected in step 4. Step 5 aims to compare the inactivation curves of the surrogate candidates with pathogens. Detailed description of the processing conditions should be provided.
6. Fitting experimental data with the linear or nonlinear models to obtain kinetic parameters that can be used for comparison of UV resistance of tested strains and final surrogate selection in process validation.
7. Comparative studies of the growth of the bacterial residuals in the post-UV-treated food product.

Following this approach, Orlowska et al. (2014) tested seven nonpathogenic *E. coli* strains: ATCC 25253, ATCC 25922, ATCC 11775, ATCC 8739, ATCC 11229, NAR, and O157 Dm3Na as potential candidates for the surrogate of pathogenic O157:H7 and non-O157 STEC strains (O111:NM, O26:H11, O145:NM, and O103:H2) in apple juice treated by UVC light at 222, 254, and 282 nm.

First, nonpathogenic strains of *E. coli* have been subjected to the high-acid stress in apple juice for 24 h. It was reported that the population of ATCC 25922, ATCC 25253, and NAR strains significantly ($P < .05$) decreased. On the contrary the population of ATCC 11229, ATCC 11775, ATCC 8739, and O157Dm3Na strains was not affected by low pH and showed acidic resistance similar to the pathogenic strains. Considering high viability in high-acid environment, *E. coli* ATCC 11229, ATCC 11775, ATCC 8739, and O157Dm3Na have been screened as three possible candidates for surrogate organism of O157:H7 and non-O157 strains in apple juice.

Next, seven nonpathogenic strains of *E. coli*: ATCC 25253, ATCC 25922, ATCC 11775, ATCC 8739, ATCC 11229, NAR, and O157 Dm3Na were screened in terms of their UV survival at 222, 253.7, and 282 nm and similar UV dose. The results have shown that the UV resistance of *E. coli* strains was spectral dependent of the used UV source. Lastly, the inactivation studies of the nine most UV-resistant strains indicated that ATCC 11229 strain could be recommended as a potential surrogate organism for UV treatments of non-O157 STEC strains at 222 nm. In the case of UVC at 254 and 282 nm *E. coli* ATCC 8739 strain met the criteria for the surrogate organism for O157:H7 and non-O157 STEC strains. In the final step, the inactivation kinetics and inactivation parameters of the most UV-resistant *E. coli* strains in apple juice were determined. The analysis of the survival curves that were nonlinear proved that the use of proper kinetic parameters is an important consideration for the selection of the surrogate organism. The inactivation rate parameter should be used in linear inactivation models. Scale parameter of nonlinear models showed a good correlation with bacterial UV resistance when Weibull distribution is used.

4.5.3.7 UVC impact on quality, composition, and sensory

As the values of UV doses in food processing have not been established by the regulators, the process has to be developed based on achieving of product specific requirements or so-called intended technical effect. This means that actual operational UV dose can vary for different categories of juices, beverages, and liquid ingredients and very much depends on product composition, absorption coefficient, rheological characteristics, initial microbial load, and reduction requirements. Also, the UV dose should be

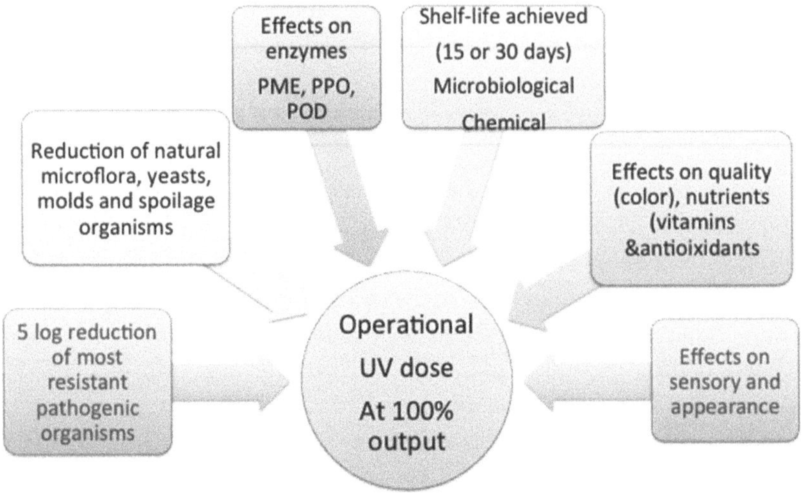

Figure 4.6 Factors juice products affecting in the establishment of operational ultraviolet dose.

optimized to achieve the best quality, nutrition, and sensory attributes as shown in Fig. 4.6. This can involve a 5-log target pathogen reduction in juices to meet HACCP, shelf-life extension of fresh juices, pasteurized milk or liquid sugars by eliminating spores or spoilage organisms.

Despite the promising status and existing regulatory approvals, the commercialization of UVC technology for preservation requires acceleration through research emphasis on the physical characteristics, quality, nutrients, enzymatic and microbial stability, and the composition changes in UVC-treated products.

Generally, the UVC treatment is not considered to pose any new nutritional safety concerns when used at the doses required to achieve 5-log reduction of target pathogenic organisms in high-acid juices and beverages. Also, the majority of studies concluded that when compared to thermal processing, UV treatment is able to better preserve the quality and nutritional attributes (Koutchma et al., 2016). However, it is known that UVC light at high doses can potentially adversely affect foods by generating free radicals by a wide variety of organic photochemical reactions. These adverse effects are very much determined by the UVC dose and in some cases the level of UVC irradiance. Depending on the UV processing dose and specific nutrient(s), UV light may have a positive, neutral, or negative effect on nutrient and enzyme retention. The key knowledge for turning the detrimental effects of UVC light into beneficial ones could lie in light-control of biological and chemical systems, such as enzymes and other compounds. To determine the most beneficial use of UV light, it is necessary to test each product for its UV spectral response. Special attention should be given to the beneficial effects of UVC light, such as increase in enzyme activity, enhancement of some antioxidants and vitamins content, and color improvement after treatment and during storage. In addition, knowledge of the degradation kinetics of vitamins, enzymes, and bioactive compounds by UVC light will allow optimization of microbial inactivation of pathogenic and spoilage organisms while minimizing losses of these health-related compounds.

Besides ensuring the absence of pathogenic microorganisms, inactivating indigenous microorganisms and enzymes that are responsible for spoilage and quality deterioration is another critical goal in food processing. For example, the quality of juice is defined by the indigenous microbial populations, indigenous enzyme activities, physical attributes (pH, Brix, color, viscosity, and cloud), chemical attributes (sugar profile, acid profile, minerals, and titratable acidity), nutritional content (ascorbic acid, folic acid, and beta-carotene), and organoleptic properties (aroma profile, flavor profile, and texture).

Shelf-life studies are commonly performed to assess the consistency of the treated juice quality throughout its targeted shelf-life. During the shelf-life study, samples are periodically tested for indigenous microorganisms, indigenous enzymes, and the evaluation of other quality attributes evaluation. In addition, factors, such as types of packaging materials (glass, polyethelen (PET), high density polyethelen (HDPE), paper carton, and pouch), storage temperature (abused and normal cold chain distribution refrigeration temperature), and transportation, should also be included in the study. Standard Plate Counting techniques are generally used to estimate the total number of viable cells in the treated juice. The yeast and mold populations and lactic acid bacteria counts are determined, as these are microorganisms that are important in juice spoilage.

Also, sensory studies should be conducted in commercial scale units to improve consumer acceptance. These data are critical for regulatory reviews of novel foods treated using UVC technology, new product and process development, and product storage stability.

4.5.3.8 Chemical and physical safety

During process validation, the production of toxic by-products, for example, and the use of unapproved additives and inadvertent introduction of potential allergens, must be avoided by proper process design and evaluation. As there is uncertainty about the production of potentially toxic substances such as free radicals or new compounds by the new process, it is essential to confirm the absence of these unwanted by-products as part of the validation program. Likewise, introduction of physical hazards, such as contamination of product by introduction of foreign objects during processing, must be prevented. Examples include introduction of glass fragments from bottles as a result of cracking through inappropriate temperature cycling, or the inclusion of metal shards from lids as a result of exceeding tolerance in filling operations. Again, properly designed and executed process validation studies will effectively consider the likelihood of such occurrences and lead to measures for their prevention, elimination, or reduction to acceptable levels.

4.5.3.9 Chemical actinometry

Chemical actinometry is a photochemical method for measuring the number of photons in a light beam integrally or per unit time. A chemical sensitive to UV light at the wavelength of interest is exposed, and the resulting photochemical changes are measured. With chemical actinometry, photochemical conversion is directly related to the number of photons absorbed. In principle, some of the requirements for an effective chemical actinometer are constant quantum yield (i.e., number of molecules of product formed per photon absorbed) over a wide range of wavelengths, high sensitivity to UV light, and compatibility with the sample matrix. Some of the most widely used chemical actinometers used to measure dose during UV processing of water are potassium ferrioxalate ($K_3Fe(C_2O_4)_3$) and potassium iodide (KI). Although these standard actinometers have been used to measure UV dose in dilute aqueous solutions, little information has been published on the use of these actinometers in high-acid solutions having large amounts of soluble solids, insoluble solids, and UV-absorbing compounds such as juices. In general, absorptivity and suspended particles, even those found in wastewater, do not approach absorbance levels encountered in juices.

Estimation of the quantum yield of actinometers provides an accurate calculation of UV dose. UV dose is defined as the delivered photons from a UV source that reacts with the actinometer resulting in a chemical change at a particular wavelength. Thus delivered UV dose (Einstein/cm^2) or (mJ/cm^2) is the ratio of the number of moles of actinometer reacted per cm^2 in relation to the quantum yield of the actinometer. The quantum yield at 254 nm can be calculated based on Eq. (4.25) (Lu, 1981)

$$Q = \frac{-dc/dt}{I_0/b(1 - 10^{-kbc})} \tag{4.25}$$

where Q is the quantum yield of the actinometer (mol/Einstein); $-dc/dt$ is the change in concentration of the actinometer with time (mol/L · s); I_0 is the incident UV intensity (mW/cm^2); b is the path length (cm); k is the molar absorption coefficient of the actinometer at 254 nm (L/mol/cm); and c is the initial concentration of actinometer dye (mol/L).

4.5.3.10 Mathematical modeling

Computing the UV fluence is another way to evaluate the performance of UV processing reactors, since software is capable of modeling UV fluence and predicting particle and fluid velocities, particle mixing, and the residence time distribution (RTD). Mathematical modeling is an essential tool in the development of a UV process for liquid foods and beverages. Modeling can be used for a number of purposes.

First, mathematical modeling can be used to predict the efficiency of microbial inactivation in a UV reactor for a specific application based on inactivation kinetics and transport phenomena.

Second, mathematical modeling of UV fluence in the reactor can assist in understanding the fluence distribution and identify the location of the least treated liquid or dead spot in the reactor. Since commercial UV reactors are of a flow-through type, they are expected to have a distribution of exposure time or RTD and fluence rate distribution resulting from UV light attenuation in a medium with high absorptive properties. The emitting characteristics of the UV light source and absorptive properties of the treated medium, the RTD in the annulus, and the annulus size and geometry will determine the UV fluence distribution in the reactor.

A further application of the mathematical modeling of UV processing is to calculate the optimal dimensions and geometry of the UV reactor for maximum inactivation performance taking into account the specific physical properties of food and the requirements of the process.

4.5.3.11 Hydraulic considerations

Flow dynamics of UV systems for the treatment of low-UVT liquids shall be evaluated for all tested products. The Reynolds (Re) number is the ratio between momentum forces (flow) and viscous forces. If the Re is high enough, then flow predominates in creating the turbulence required for adequate surface refreshment and mixing. Re numbers then shall be calculated as

$$\mathrm{Re} = v d \rho / \mu \tag{4.26}$$

where d (m) is the characteristic dimension, ρ is the density of fluid (kg/m^3), and μ is the dynamic viscosity (Pa s). Velocity (v) (m/s) is calculated as ($v = Q/A$), where Q is a volumetric flow rate (m^3/s) and A (m^2) is a cross-section area of the tube.

The Re number shall be calculated for the range of flow rates and for each product if viscosity differs and for a range of processing temperatures used. For instance, as it was reported by Koutchma et al. (2007), the Re numbers were calculated for all tested groups of tropical juices whose viscosities varied in a wide range. It was found that the magnitude of Re for pineapple and orange juice (known as non-Newtonian liquids) was less than 2000, indicating that at the selected flow rates, the hydraulic regime was laminar. However, for juices characterized as Newtonian liquids (lilikoi, guava, apple, and watermelon) at the flow range from 3.5 to 7 gpm used in the trails, the flow dynamics could be characterized as a turbulent flow with $\mathrm{Re} > 2000$.

Another important feature of hydraulics in the coil geometry UV system concerns the development of secondary flows in the tube. It has been demonstrated by Dean (1927) that a secondary flow field accompanies laminar flow of fluid through a coiled tube. The curvature radius that exists between the inner and outer boundary layer causes the phenomenon called the Dean effect. The Dean number, De, is the unique similarity parameter governing fluid motion in such a flow configuration. The Dean number is calculated a $\mathrm{De} = \mathrm{Re}\sqrt{D/D_C}$, where D (m) is the tube diameter and D_C (m) is the coil diameter. Secondary flow eddies are highly promoted within the range of D/D_C of 0.03–0.10. The magnitude of Dean numbers from 65 to 132 for pineapple and orange juices, and from 573 to 1145 for apple and lillikoi, and ratio of $D/D_C = 0.046$, indicated that secondary flows developed in juices providing additional mixing in case of both laminar and turbulent flow regimes.

A stimulus–response technique can be used for the experimental determination of RTD and UV fluence distribution in the UV systems (Koutchma and Parisi, 2004). Koutchma et al. (2007) used a 50% caramel solution as a stimulus, or tracer in the validation studies of the coiled tube UV system. The special loop in the inlet tube of the reactor containing an inlet port and two-way valves was built up to inject the tracer into the reactor tube. A 50% caramel model solution was injected into the center of the loop assembly when the valves were closed. When flow achieved a steady-state condition, the valves were opened and the caramel solution was mixed with the water stream in the tube. The change in the caramel concentration was measured with time at the output. The distribution of caramel concentration or $c(t)$ curve reflected the RTD in the UV reactor. The normalized curve $E(t)$, which represents the fraction of fluid leaving the system at each time or RTD function can be constructed from the $c(t)$ curve using the summation technique (Eq. 4.27):

$$E(t) = \frac{c(t)}{\int_0^t c(t)dt} \tag{4.27}$$

The mean residence time of the flow $\bar{t}$ is calculated from the function $E(t)$ and time t using the summation technique (Eq. 4.28):

$$\bar{t} = \int_o^t tE(t)dt \tag{4.28}$$

The theoretical residence time ($t = V/Q$) can also be calculated from the fill volume of the reactor V (m^3) and volumetric flow rate Q (m^3/s). The ratio of $\bar{t}/t$ is used as indicator whether full use is made of the reactor volume. If the value is less than 1, it indicates that the effective volume is much less than the actual volume of the reactor (Downey et al., 1998).

4.5.3.12 Design of commercial UVC systems

Due to low UVT and viscosity challenges, novel UVC systems for low-UVT and opaque beverages have to be specifically developed and differ from designs normally employed for water treatment. Thin film laminar, annular turbulent, static, and dynamic mixing regimes are used to effectively treat juice suspensions, milk products, and beverages. Currently, only a few designs of continuous flow UV apparatus are available for processing various types of these products. Some of these UVC systems are under development or testing and only a few are commercially available. Limited information has been reported on the application of commercial scale UVC systems, including CiderSure (thin film laminar flow; FPE, NY, United States), Salcor module (Dean flow in coiled tube, turbulent; CA, United States), SurePure (thin film, turbulent; SurePure Inc., Zug, Switzerland), and AseptoRay (static mixer, turbulent, MGT, Israel).

Food processors who are interested in using UVC technology have to select the most appropriate system for their application and establish processing conditions to achieve their safety and shelf-life objectives. In the same time processing conditions have to provide the required level of quality while retaining nutritive constituents, product sensory, and structure attributes.

4.5.4 Validation of commercial UV equipment and process controls

The key elements of equipment validation include initial inspection, installation and operational qualification, performance qualification, and commissioning. Prior to the installation of the equipment, a description of the equipment, checklist of the software, and accurate supporting documentation must be provided. The manufacturer of UV reactors must provide information on the proper installation and operation of the reactor. This includes performance characteristics such as nominal pressure, flow rates, head loss, pressure rating of the sleeve, UV intensity sensor, and monitoring window used; assembly and installation instructions; electrical requirements; and operation and maintenance manuals. During installation and operational qualification, sensors need to be calibrated against standards (e.g., National Institute of Standards) and control loops have to be tuned.

Key aspects of the validation of UV systems for the treatment of liquid food, juices, and beverages include:

- Documentation describing all aspects of the UV reactor that impact the relationship between operating UV dose and microbial inactivation;
- Information on the range of flow rates, lamp output power, absorptive and physico-chemical properties of fluid being treated and their variation;
- Data relating inactivation of a challenge microorganism and/or its surrogate and the associated UV dose;
- Interpretation of validation results to ensure the UV system meets performance requirements.

UV systems undergoing validation must comply with this documentation.

4.5.4.1 Critical process parameters and ways of their monitoring

UVT, flow rate/product viscosity, process temperature, and UV intensity are critical parameters that will affect UV system performance requirements. A relationship between the required UV dose and these parameters must be established and then monitored to ensure the disinfection of target organisms.

4.5.4.2 UV transmittance

UVT has a strong effect on the dose delivery of a UV reactor. As UVT decreases, the intensity throughout the reactor decreases, which reduces the dose the reactor delivers. UV reactors are typically sized to deliver the required UV dose under specified UVT conditions for the application. Design UVT is the minimum UVT that will typically occur at the design flow. The design UVT and design flow are typically used by the UV manufacturer to determine the appropriate UV equipment for a target level of inactivation. The design UVT may not necessarily be the minimum operating UVT.

Minimum operating UVT is the lowest UVT expected to occur during lifetime of the UV facility. Understanding the minimum UVT is critical because the UV reactor should be designed and validated for the range of UVT and flow rate combinations expected to avoid off-specification operation. Quantifying both a design UVT and the full range of UVT expected during operation is essential because the UV reactor should be validated for the range of UVT and flow combinations expected to avoid off-specification operation.

UVT measurement and data analysis can be used to determine design UVT and UVT range. The effect of product type, quality, including concentration of suspended particles or added ingredients on turbidity, and UVT has to be determined. UVT data collection and measurements may be necessary from locations that are representative of the potential facility location. If UVT data are not available, weekly UVT measurement is recommended. UVT readings can be taken using an online monitor or samples can be collected manually and analyzed. UVT monitors may be included with the UV system, depending on the control concept used. A UVT monitor measures the transmittance of the product flowing through the reactor. A UVT monitor may consist of multiple UV intensity sensors located at various distances from a UV lamp. The differences between the intensity readings are interpolated to determine the UVT of the product.

4.5.4.3 Turbidity

Particle content, association, and clumping of microorganisms can also affect UV performance. Particles can absorb and scatter UV light to varying degrees based on their size and composition.

4.5.4.4 Flow rate and pattern

The UV facility design criteria should identify the average, maximum, and minimum flow rates that the UV reactors will experience. Design flow rate is the maximum flow that can be treated at the UV facility.

4.5.4.5 Fluid dynamics and Reynolds number

The flow rate/pattern strongly influences fluid dynamics and the total absorbed UV dose because the position and the residence time of the microorganisms in certain regions of the irradiance field can vary. Flow characteristics and regimens have to be evaluated for UV processed liquid products, and the Re number is calculated according to Eq. (4.24). The Re number can be used to determine if flow is laminar, transient, or turbulent. The flow is laminar—when $Re < 2000$; transient—when $2000 < Re < 4000$; and turbulent—when $Re > 4000$.

Viscosity of liquid strongly affect Re number and is a function of temperature meaning that Re number shall be evaluated for a range of processing temperatures.

The analysis of Re numbers has to be conducted to determine the Re numbers in the UV system and flow regimes at the design, average, maximum, and minimum processing flow rates.

4.5.4.6 Pressure

Maximum operating pressure is recommended by the UV system provider and should be monitored and reported during UV system operation.

4.5.4.7 Process and product temperature

The processing temperature can affect the viscosity of the product, mixing, and efficiency of the LPM lamp operation. The surface temperature of LPM lamp is roughly 40°C, LPA lamp is around 90°C. The UV unit design criteria should identify the average, maximum, and minimum temperatures that the UV reactors will experience. Preheating viscous liquids may lower the viscosity but be within the optimal lamp operation temperature. The inlet and outlet temperatures of product should be controlled by the built-in thermocouples.

4.5.4.8 UV lamp intensity

Online monitoring and control strategies for UV intensity set point require measurements of:

- UV intensity,

- flow rate,
- lamp status,
- liquid temperature, and
- UVT.

4.5.4.9 UV sensors

UV reactors are equipped with at least one online UV intensity sensor that measures the UV intensity at some point within the UV reactor. Measurements made by the online UV intensity sensors are used to ensure that the reactor is delivering a UV dose that meets the regulatory requirement. For example, the CiderSure 1500 UV reactor is equipped with two sensors. Based on sensor readings at two points within the flow path an "auto" setting can adjust flow using an algorithm based on the transmittance of the UV light through the treated fluid.

The UV intensity sensor should detect germicidal UV radiation and produce a standardized output signal (4–20 mA) proportional to the UV intensity incident on the sensor. UV intensity sensors must be calibrated to an absolute intensity standard and have a suitable measurement range, angular response, wavelength response, linearity, and stability for monitoring and controlling UV dose delivery by the UV reactor. An ideal UV intensity sensor has the following characteristics:

1. Responds linearly as the UV intensity within the reactor changes over the measurement range of the UV intensity sensor.
2. Responds independent of product temperature.
3. Stable over time.
4. Has a fixed acceptance angle.
5. Responds only to germicidal UV ranges.

The manufacturer of the UV reactor must provide documentation on the UV intensity sensor and monitoring window that is separate from the UV sensor. A separate monitoring window provides reproducible measurements of UV intensity by the online and reference UV intensity sensor.

The measurement uncertainty of a UV sensor is due to the uncertainty in UV intensity sensor calibration, linearity over the working range, and stability over time and temperature. In addition, UV sensor degradation will occur due to exposure to UV light. The reference UV intensity sensor should not be exposed to UV light longer that it takes to make the measurement. Alarms, interlocks, equipment communication protocols, utility cleaning, and consumable requirement should also be verified at the installation stage. Analytical methods for in-process controls and release testing have to be fully validated according to General Manufacturing Practices (GMPs) and current standards.

4.5.4.10 Validation of commercial UV systems for low-UV transmittance liquid products

The objective of commercial UV system validation is to determine the operating conditions under which a UV reactor delivers the validated dose.

- The validated dose must be greater than or equal to the required dose to receive log inactivation credit for a target organism.
- Validation testing also establishes the operational set points used during reactor operations to confirm delivery of the validated dose.

Recommended validation steps include:

1. Detailed UV reactor documentation
   a. Describes the wetted dimensions of the reactor and optical properties of the lamps, sleeves, UV sensors, and UV sensor ports that impact UV dose delivery and monitoring;
   b. UV system inlet and outlet piping with dimensions.
2. Validation test plan
   a. Description of the test sequence;
   b. Challenge microorganism stock solution preparation and enumeration;
   c. Third-party oversight;
   d. Product physico-chemical characteristics and quality measurement methods (pH, natural microbiota, UVT, turbidity units (NTU), color, nutrients, sensory, etc.);
   e. Functional test methods (headloss, power, UV sensor reference checks).
3. Results of CB testing
   a. Microbial suspension (volume, depth, concentration), irradiance measurement, calculations of Petri factor, and dose.
4. Results of full-scale commercial testing
   a. Flow rate, UV intensity, UVT, Lamps power, status;
   b. Inlet and outlet concentration of challenge organism;
   c. Quality control and checks of microbial suspension; radiometer and other instruments;
   d. Calculation of the validated dose (RED), log inactivation credit, and validated operating conditions (flow, lamps status, UVT);

e. Measuring UV dose effects on physico-chemical parameters (Brix, pH, viscosity), quality (color), and sensory parameters (odors, flavor);
f. UV dose effects on formation chemical compounds (Hydroxymethylfurfural (HMF), furan).

4.6 Conclusion

According to the reported effects on pathogenic and natural microbiota UVC treatment of opaque beverages and liquids can potentially provide safety levels and shelf-lives of products during refrigerated storage similar to that for products treated with HHP with no effects on the beverages' essential quality, nutritional and sensory attributes. The lower initial investment, operating costs, packaging flexibility, and size requirements associated with industrial UVC treatment show the feasibility of this technology as the primary nonthermal processing alternative for the premium beverages category. Due to the limited data of UVC resistance of foodborne pathogens and absence of common established practices to measure and evaluate UVC inactivation kinetics in opaque beverages products, the challenge remains to properly establish UVC preservation process, validate the process, and obtain regulatory approvals. Because the UVC dose has not been established by the regulators, individual UVC processes have to be developed and validated based on achieving product-specific requirements. Therefore the main challenge for food processors interested in using UV technology is to select the most appropriate UVC system for their application and establish processing conditions to achieve their safety, shelf-life, and quality objectives. The further development of validation and verification practices such as phosphatase test in milk pasteurization, optimization of UV equipment, and further regulatory approvals can certainly bring the UVC technology for beverages closer to commercial implementation.

Also, not all of the issues of validation of UV technology for juices and other liquid foods were discussed in this chapter, but a step forward was made in understanding the needs and important steps for the future.

References

Anonymous, 2000. Kinetics of microbial inactivation for alternative food processing technologies. Institute of Food Technologists (Journal of Food Science Supplement). <http://vm.cfsan.fda.gov/~comm/ift-pref.html>.

Beauchamp, S., Lacroix, M., 2012. Resistance of the genome of Escherichia coli and *Listeria monocytogenes* to irradiation evaluated by the induction of cyclobutane pyrimidine dimers and 6–4 photoproducts using gamma and UV-C radiations. Radiat. Phys. Chem. 81 (8), 1193–1197.

Bolton, J.R., Linden, K.G., 2003. Standardization of methods for fluence (UV dose) determination in bench scale UV experiments. J. Environ. Eng. 129 (3), 209–215.

Chun, H.H., Kim, J.Y., Lee, B.D., Yu, D.J., Song, K.B., 2010. Effect of UV-C irradiation on the inactivation of inoculated pathogens and quality of chicken breasts during storage. Food Control 1, 276–278.

Dean, W.R., 1927. Philos. Mag. J. Sci. 4, 208–223.

Downey, D., Giles, D., Delwiche, M., 1998. Finite element analysis of particle and liquid flow through an ultraviolet reactor. Comput. Electron. Agric. 21, 81–105.

EFSA, 2016. Safety of UV-treated milk as a novel food pursuant to Regulation (EC) No. 258/97. EFSA J. 14 (1), 4370. <http://www.efsa.europa.eu/en/efsajournal/pub/4370>.

EFSA, 2018. EU authorized list of UV treated novel foods. <https://ec.europa.eu/food/safety/novel_food/authorisations/list_authorisations_en>.

European Union, 1997, 258/97.

Gomez-Lopez, V.M., Ragaerta, P., Debeverea, J., Devlieghere, F., 2007. Pulsed light for food decontamination: a review. Trends Food Sci. Technol. 18, 464–473.

Green, A., et al., 2018. Inactivation of *Escherichia coli, Listeria and Salmonella* by single and multiple wavelength ultraviolet-light emitting diodes. Innov. Food Sci. Emerg. Technol. 47, 353–361.

Harm, W., 1980. Biological Effects of Ultraviolet Radiation. Cambridge University Press.

Health Canada, 2004. Ultraviolet light treatment of apple juice/cider using the CiderSure 3500. Novel Food Information. <http://www.hc-sc.gc.ca/fn-an/gmf-agm/appro/dec85_rev_nl3_e.html> (accessed 29.02.08.).

Koutchma, T., Parisi, B., 2004. Biodosimetry of *E. coli* UV inactivation in model juices with regard to dose and RTD distribution in annular UV reactor. J. Food Sci 69 (1), 14–22.

Koutchma, T., Parisi, B., Patazca, E., 2007. Validation of UV coiled tube reactor for fresh fruit juices. J. Environ. Sci. Eng. 6, 319–328.

Koutchma, T., Popovic, V., Ros-Polski, V., Popielarz, A., 2016. Effects of ultraviolet light and high-pressure processing on quality and health-related constituents of fresh juice products: UV & HPP effects on juice quality and health related constituents. Compr. Rev. Food Sci. Food Saf . Available from: https://doi.org/10.1111/1541-4337.12214.

Kuo, J., Chen, C., Nellor, M., 2003. Standardized collimated beam testing protocol for water wastewater ultraviolet disinfection. J. Environ. Eng. pp. 773–779.

Lu, Y.C., 1981. The determination of the quantum yield of benzoquinone by a modified method of actinometry. Ph.D. Thesis, University of Cincinnati, pp. 1–46.

Orlowska, M., Koutchma, T., Kostryanska, M., Tang, J., 2014. Surrogate organisms for pathogenic O157:H7 and non-O157 Escherichia coli strains for apple juice treatments by UV-C light at three monochromatic wavelengths. Food Control 47, 647–655.

Qualls, R.G., Flynn, M.P., Johnson, J.D., 1983. The role of suspended particles in ultraviolet disinfection. J. WPCF 55, 1280–1285.

Regulation (EU) 2015/2283 of the European Parliament and of the Council of November 25, 2015.

Sommers, C.H., Geveke, D.J., S. Pulsfus, S., Lemmenes, B., B., 2009. Inactivation of *Listeria innocua* on frankfurters by ultraviolet light and flash pasteurization. J. Food Sci. 74 (3).

Sommers, C.H., Scullen, O.J., Sites, J., 2010. Inactivation of foodborne pathogens on frankfurters using ultraviolet light (254 nm) and GRAS antimicrobials. J. Food Saf. 31, 1–12.

U.S. Food and Drug Administration, 2001. 21 CFR Part 179. Irradiation in the production, processing and handling of food. Fed. Regist. 65, 71056–71058.

US FDA, 2001. Hazard Analysis and Critical Control Point (HACCP): Procedures for the Safe and Sanitary Processing and Importing of Juice. Final Rule. Federal Register, 66, No. 13. U.S. Food and Drug Administration, Washington, DC.

U.S. Food and Drug Administration, 2012. 21 CFR Part 172. § 172.381 Vitamin D2 bakers' yeast. Federal Register/Vol. 77, No. 168/Wednesday, August 29, 2012/Rules and Regulations 52229. Food Additives Permitted for Direct Addition to Food for Human Consumption; Vitamin D2 Bakers Yeast.

CHAPTER 5

Case study of validation of ultraviolet light pasteurization of sugar syrups

Tatiana Koutchma
Novel Food Processing Technologies, Agriculture and Agri Food Canada, Guelph, ON, Canada

5.1 Evaluation of ultraviolet dose in commercial ultraviolet system

First, the study was conducted to quantify the inactivation of *Bacillus subtilis* spores artificially inoculated into two types of sugar syrups, high fructose corn syrup (HFCS) and liquid sucrose, by exposure to ultraviolet (UV) light at 253.7 nm in a commercial-scale reactor. Syrups were inoculated with 10^5 CFU/mL *B. subtilis* spores and pumped through a UV reactor at various flow rates. *B. subtilis* levels were quantified prior to treatment (untreated control) and after UV processing (treated) to determine the level of inactivation achieved. Approximately $5\log_{10}$ reduction of *B. subtilis* spores was observed in the liquid sucrose at the fastest flow rate of 35.7 gpm. To quantify "dose—response" relationships of spore inactivation, experiments were conducted with the single central lamp at three selected flow rates and ambient temperature. UV dose delivered in the UV reactor was evaluated using biodosimetry approach and mathematical modeling of fluence rate distribution and residence time distribution (RTD).

Objectives of the study are

1. to determine the level of inactivation of *B. subtilis* spores in sugar syrups exposed to UV light in the commercial unit and
2. to evaluate the UV light dose delivered by the commercial-scale UV light equipment in sugar syrup.

5.1.1 Experimental approach

The schematic diagram of the experimental setup of UV treatment of liquid sucrose with commercial UV light unit is shown in Fig. 5.1 along with the legend of other supplementary equipment. Sucrose samples were

Validation of Food Preservation Processes based on Novel Technologies
DOI: https://doi.org/10.1016/B978-0-12-815888-3.00004-X

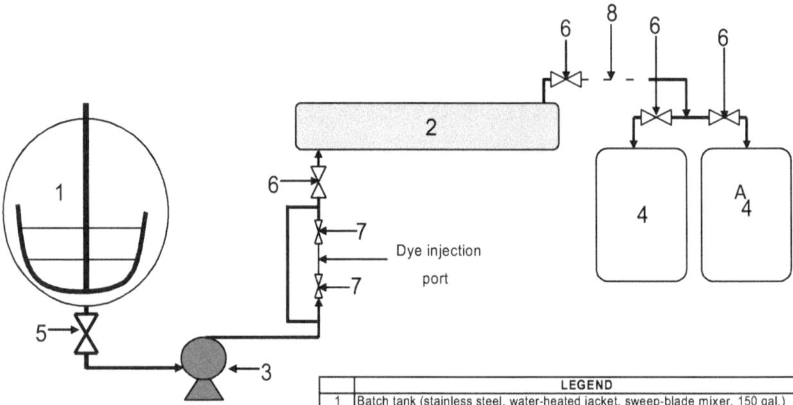

	LEGEND
1	Batch tank (stainless steel, water-heated jacket, sweep-blade mixer, 150 gal.)
2	American Ultraviolet UV reactor
3	Reeves rotary positive-displacement pump (5 hp, stainless steel rotors)
4	Product receiving tank (stainless steel, 50 gal.)
5	2" ball valve (tri-clover connection)
6	1.5" butterfly valve (tri-clover connection)
7	0.5" butterfly valve (tri-clover connection)
8	Dye testing sightglass (1" polymer tubing, tri-clover connection)
—	0.5" stainless steel pipe (tri-clover connection)
—	1.5" stainless steel pipe (tri-clover connection)
—	2" fiberglass tubing (tri-clover connection)

Figure 5.1 Scheme of the experimental setup of the UV unit.

inoculated in the inlet tank with a sweep-blade mixer (#1). After warming up the UV unit for 10 min, the sample was pumped using rotary displacement pump (#3, Reliance Electric, Greenville, SC) through the UV reactor (#2, American Ultraviolet Co., Lebanon, MI) to the product receiving tank (#4). Sucrose entered one side of UV module through a stainless-steel tube and then exited through an outlet tube. The process was run for 1 min before treated samples were pulled. For the first minute, all product ran through the UV reactor was collected in the first tank 4. At the end of first minute, the syrup was diverted to a second tank, designated tank 4A. Sucrose was sampled before UV exposure and immediately after a single pass through the reactor. Two approaches were used for sampling after UV treatment. A 40–50-mL aliquot of treated syrup was aseptically removed at the end of the UV reactor and transferred to a sterile centrifuge tube. Treated samples were pulled at the beginning, middle, and end of each run to determine if there is variability across the processing run. The beginning sample was always taken after 1 min or when a steady-state condition of the flow was achieved. In addition, volume averaged samples were pulled from the output tank (tank 4A) at the end of the process run, in triplicate. Prior to and following each treatment session, the UV reactor was cleaned and sanitized according to the CIP procedure.

5.1.1.1 UV light treatment unit

Commercial-scale continuous flow UV unit equipped with seven low-pressure amalgam lamps was validated for treatment of liquid sugar syrups. Each lamp was installed in quartz sleeve and product flows in the gap between lamps and chamber walls. The technical and geometrical characteristics of the UV unit are specified in Table 5.1.

5.1.1.2 UV lamps

The lamps were 320-W output power low-pressure, high-intensity amalgam lamps. According to the manufacturer, 90% of UV light output is within the wavelength of 233.7−273.7 nm. The UV light output of the lamp is approximately 105−110 W and produced 750 μW/cm^2, which is a 35% efficiency rating. The lamp life time was 9000 h. The lamp arc length is 149 cm 58.66″ with the OD of 19 mm. Lamps sleeves are type 214 clear fused quartz circular tubing rated for the UV transmission of 94%.

5.1.1.3 UV light unit experimental configurations

The microbial inactivation experiments were conducted after one pass through the UV reactor. In the initial validation tests, it was found that the UV fluence and inactivation efficiency of the tested UV lamps were relatively high, and there were no spore survivors when all seven lamps were turned "ON" indication larger than 5$\log_{10}$ microbial reduction. To enumerate spore counts after UV treatment and to evaluate UV fluence accurately, the experiments were conducted at three selected flow rates with four experimental configurations of the UV lamps as shown in Fig. 5.2.

1. Three side lamps "ON" (#1, 3, and 5 or #2, 4, and 6);
2. The single peripheral lamp #4 "ON";

Table 5.1 UV reactor technical characteristics.

1. Number of chambers	1	
2. Peak flow rate	30 gpm	
3. Number of lamps	7	
4. Vessel diameter	6″	15.24 cm
5. Laying length	72″	
6. Working length (L)	56.25″	142.88 cm
7. ID of the reactor	5.87″	14.91 cm
8. Total cross-sectional area of the reactor	27.05″	174.51 cm
9. Total volume of UV reactor V_r	1521.49″	24,932.71 cm^3
10. OD of quartz sleeve	1.42″	3.61 cm

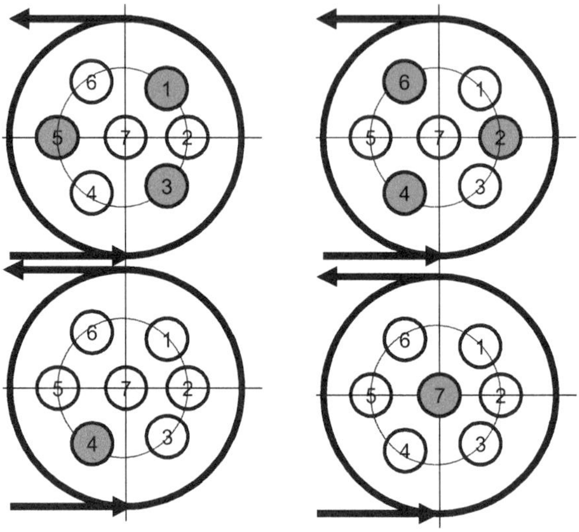

Figure 5.2 Configurations of UV lamps in the UV unit used for testing.

3. The single central lamp #7 "ON"; and
4. All seven lamps "ON" (not shown).

5.1.1.4 Test products

Liquid sucrose was received from Savannah Foods & Industry (Port Wentworth, GA). Samples were taken for physical parameters testing from each lot. HFCS was received from Cargill Corp (Eddyville, IA).

5.1.1.5 Physical and optical characteristic measurements

The physical properties of liquid sucrose and HFCS, including density, viscosity, total solid weight fraction, absorption coefficient, turbidity, pH, and degree Brix, have been measured to develop a UV light process that includes calculation of UV fluence absorbed by the medium and to evaluate products' flow dynamics and flow parameters.

Average bulk density was determined by weighing a known volume amount of sample. The measurements were repeated with three samples at 20°C and checked with water, for which density at this temperature is known.

Viscosity was measured using a Brookfield LVDV-II + type viscometer (Brookfield Engineering, Middleboro, MA), calibrated at 20°C with distilled water and standard fluid. Viscosity was measured in the range of temperatures from 68 to 100°F.

Soluble solids content was measured using the Abbe refractometer (American Optical Corporation, Buffalo, NY).

pH was measured using an Orion 520A pH meter (Fisher Scientific, Hanover Park, IL) at room temperature (RT).

Absorbance of sucrose syrup and HFCS was measured at 254 nm using double-beam spectrophotometers (Caray 100 by Varian, Walnut Creek, CA; Lambda 2 by Perkin Elmer, Boston, MA). Due to the low light transmittance, sample solutions were evaluated using matched demountable fused quartz cuvettes (NSG Precision Cells Inc., NJ) with path lengths of 0.1, 0.2, 0.5, and 1.0 mm. The absorption coefficient of the sample solution was determined from the slope of the absorbance versus path length.

Turbidity was measured using an HACH 2100N Turbidimeter (Hach Company, Loveland, CO).

Color was measured using a Hunter Labscan XE Colorimeter (Hunter).

All measurements were made in triplicate.

5.1.1.6 Inoculum preparation

B. subtilis ATCC 6633 was received from the American Type Culture Collection and used in all inoculated pilot plant experiments. This microorganism was selected because the UV inactivation response of *B. subtilis* spores in water has been well documented, it is more resistant to UV than vegetative bacteria, is widely used in validating water and wastewater UV treatments, and is easy to analyze (Chang et al., 1985; Nicholson and Galeano, 2003; Sommer and Cabaj, 1993; Sommer et al., 1995, 1998).

Sterile tryptic soy broth (TSB; 0.5 mL), was added to the freeze dried culture. The entire volume was then transferred to TSB (10 mL) and incubated with shaking at 30°C for 18–24 h (American Type Culture Collection, 2005). An aliquot (100 μL) of the overnight culture was transferred to TSB (10 mL) and incubated at 30°C for 72 h. One milliliter of the *B. subtilis* culture was inoculated into each of four flasks containing 125 mL TSB and incubated at 30°C for 18–24 h in a shaker incubator set at 80 rpm (Model 12400, New Brunswick Scientific, Edison, NJ). Five hundred nutrient agar plates containing yeast extract and manganese sulfate (NAYEMn) were inoculated with the overnight culture (1 mL) and incubated at 37°C for up to 7 days until 90%–95% sporulation was observed (Claus and Berkeley, 1986; Hamouda et al., 2002). Spores were harvested by scraping the plates and suspending in sterile 50% ethanol. The resulting suspension was incubated at ambient temperature (25°C–26°C) on an orbital shaker overnight to lyse vegetative cells. The overnight suspension was centrifuged (Model 5804R,

Eppendorf, Hamburg, Germany) at 2500 × *g* for 10 min. Two washing steps were conducted using cold, sterile deionized water and centrifugation as described above (Hamouda et al., 2002). This process resulted in 1200 mL of spore crop available for the pilot plant experiments.

A total count and a heat shock count were conducted to determine the level of spores in the crop. Heat shock count was determined by heating 3 mL of spore crop to 80°C in a circulating water bath (RC20, MGW Lauda, Western Germany) and holding for 10 min. Heated and unheated samples were serially diluted in 0.1% peptone water (PW) and spread plated on tryptic soy agar (TSA). Plates were incubated at 35°C for 18–24 h before enumeration.

5.1.1.7 Preliminary experiments to determine stability of spores in syrups

Preliminary experiments were conducted in the laboratory to determine if *B. subtilis* spores could be successfully recovered from artificially inoculated sugar syrups using standard methods. For each test run, 200 μL of spore suspension was added to c. 1000 mL of prewarmed syrup to achieve an initial target inoculum level of c. 105 CFU/mL of syrup. Inoculated syrup was mixed for 20 min by mechanical stirring, to mimic pilot plant conditions. After mixing, inoculated syrup was held at ambient temperature for a period of 2 h, to replicate conditions from time of inoculation to actual laboratory analysis during the pilot-scale studies. Dilutions of the inoculated syrup were prepared in 90-mL volumes of syrup and selected dilutions were spread plated on TSA. Plates were incubated at 37°C for 20–24 h.

5.1.1.8 Product inoculation

For each test run, an aliquot of spore suspension (Table 5.2) was added to c. 500 mL of prewarmed syrup, which was pulled directly from the mixing tank, and mixed by stirring with a sterile pipet. The inoculated syrup

Table 5.2 Processing parameters of liquid syrups.

Product	Flow rate (gpm)	Product volume (gal)	Inoculum volume (mL)
Sucrose	35.7	75	75
	7.6	24	24
	3.3	20	20
Fructose	15	45	45
	1.5	20	20

was added to the prewarmed syrup in the mixing tank, to obtain a target inoculum level of 10^5 CFU/mL of syrup. Inoculated syrup was mixed in the covered mixing tank for a minimum of 20 min to ensure even distribution of spores in the syrup.

5.1.1.9 UV treatment

Various flow rates in the range from 1.5 to 35.7 gpm were investigated for each product type (Table 5.2). The initial target inoculum level was 10^5 CFU/mL syrup. Treated samples were aseptically removed after one pass for microbial analysis. Each experiment was conducted in triplicate. Product temperature for all experiments was 80°F. UV lamps usage and configuration were dependent on the product type and flow rate.

5.1.1.10 Sampling

Triplicate samples of inoculated, untreated syrup were aseptically removed from the mixing tank and transferred to sterile sample containers. Two sets of treated samples were pulled. Volume averaged samples were pulled from the output tank at the end of the process run, in triplicate. Process control samples, representing the beginning, middle, and end of each process run, were sampled directly from the output stream into the output tank, in duplicate. All samples were transported back to the laboratory and all analyses were conducted within 2 h of sampling.

5.1.1.11 Enumeration of B. subtilis

B. subtilis levels were quantified for treated and untreated samples. Serial dilutions were prepared in PW, spread plated onto TSA plates, and incubated at 35°C for 24 h. Initially, plates were held at 35°C for up to 7 days, to allow for the recovery of injured spores that may require additional incubation time to produce visible colonies. No diffrences in spores plate counts were observed on any of the plates when comparing 24 h to 7 days of incubation, therefore for all subsequent experimental plates were incubated for 24 h prior to enumeration.

When low levels of survivors were expected, a filtering procedure was utilized. The 10- and 100-mL aliquots of syrup were diluted 1:10 in sterile diluent and passed through a 0.22-μm nitrocellulose filter. The filter was aseptically removed from the filtering apparatus and placed onto the surface of a TSA plate containing tetrazolium chloride (TTC). Plates were incubated at 37°C for 18–24 h. Colonies appeared pink on the plate due to the addition of TTC to the growth media.

5.1.1.12 Fastest moving particle studies

A stimulus–response technique was used for experimental determination of RTD of liquid syrup volume. Fluorescein disodium salt dehydrate, 90% dye (Acros Organics, Fair Lawn, NJ), was used as a tracer to measure minimum residence time in the UV reactors at the four selected flow rates of 35.7, 20.0, 7.6, and 3.3 gpm. The fluorescein disodium salt dehydrate is a concentrated dye used primarily to measure residence time in rivers, aqueducts, glaciers, and plumbing systems. The dye colored the syrup without significantly altering the viscosity. The reactor source tank was filled with an appropriate volume of liquid sucrose, depending on the desired flow rate. A special loop in the inlet tube of the reactor containing inlet port and two-way valves was built up to inject the tracer into the reactor tube. The scheme of the injecting device is shown in Fig. 5.3 and contained three two-way valves.

The 35 cc of the fluorescein disodium stock solution was injected into the center of the loop assembly when valves were closed. When flow achieved a steady-state condition, the valves were opened and the dye solution was mixed with the syrup stream in the tube. As soon as the valve was opened, the sight glass near the reactor outlet was observed for a change in color. The time in which the sample turned green was noted as the minimum residence time.

5.1.2 Results

5.1.2.1 Absorptive and physical properties of liquid sucrose and HFCS

Average bulk density, viscosity, absorbance, turbidity, soluble solids content, pH, water activity, and color were measured at RT and summarized in Tables 5.3–5.5.

The measured values of physical parameters were in close proximity with numbers provided by the manufacturer. Viscosity of HFCS was higher than that of liquid sucrose in the range of 60°F–100°F. The viscosity of liquid sucrose decreased with increase in temperature from

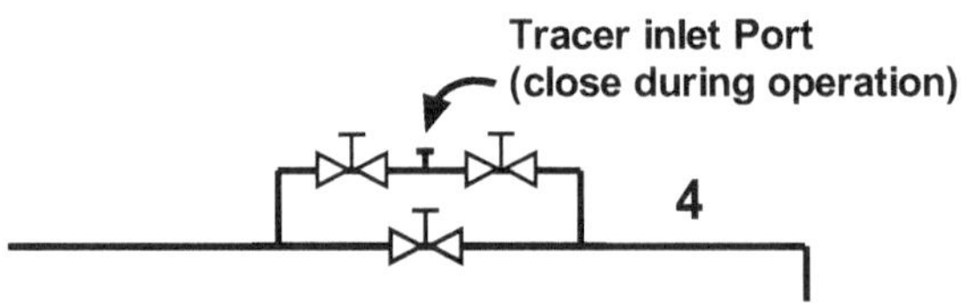

Figure 5.3 Tracer inlet port to inject dye.

Table 5.3 Physical properties of liquid sucrose.

Property	Units	Measured	Provided
Density	g/mL	1.35	1.3281
Viscosity	cP		
60°F			250
68°F			200
80°F		186.8 ± 2.47	100
90°F		90.5 ± 0.71	
100°F		53.3 ± 2.47	60–100
pH		5.93	6.5–8.5
Brix		66.9	67.5
Water activity		0.856	0.851
Dry basis	%		97

Table 5.4 Absorptive properties of liquid sucrose.

Property	Units	Measured	Provided
Absorption coefficient	cm^{-1}	0.421	
Turbidity	NTU	4.75	
Color	L, a, b	4,29; 0.48; 0.14	35 (RBU max)

Table 5.5 Physical and absorptive properties of liquid fructose (HFCS).

Property	Units	Measured	Provided
Density	g/mL	1.40	1.37
Viscosity	cP		
60°F		–	–
68°F		–	1520
80°F		655.9	760
90°F		268.7	520
100°F		133.8	360
pH		3.57	3.3–4.3
Brix		76.88	76.5–77.5
Water activity		0.67	0.65
Absorption coefficient	cm^{-1}	2.99	–
Turbidity	NTU	0.388	

186 cP at 80°F to 53.3 cP at 100°F. A difference in pH and a_w between the sucrose syrup and the HFCS was also observed (Tables 5.3 and 5.5). The absorption coefficient of liquid sucrose was approximately seven times lower than that of HFCS.

5.1.2.2 Flow dynamics

Flow dynamics was evaluated for the liquid sucrose at the entrance section of the UV reactor and in the gap between lamps. Reynolds (Re) numbers were calculated as ($\mathrm{Re} = vd\rho/\mu$) based on the measured flow rate at fluid temperature of 80°F and 100°F, where d is the characteristic dimension, ρ is the density of fluid, and μ is the dynamic viscosity. To calculate the Re number in a gap, the hydraulic diameter h was used as a characteristic dimension $h = d$. The hydraulic diameter, h, is used instead of the geometrical diameter for channels of noncircular shape. Hydraulic diameter h is defined as $h = 4\,A/P$, where A is a cross-sectional area between lamps ($A = 60.06\ \mathrm{cm}^2$), P is a wetted perimeter ($P = 8.73\ \mathrm{cm}^2$). Velocity v in the gap was calculated as $v = Q/A_{\mathrm{wa}}$, where Q is a volumetric flow rate and A_{wa} is a wetted cross-sectional area. The geometric parameters used in the calculations are summarized in Table 5.6.

The calculated Re numbers for the flow rate used in the study in the entrance pipe and in the gap between UV lamps are summarized in Tables 5.7 and 5.8. The magnitude of the Reynolds number (less than 2000) indicates that at the selected flow rates, the hydraulic regime in the UV reactor was laminar. This means that due to laminar flow, the fastest flowing particle in liquid sucrose travels twice as fast as the average particle. The residence time for the fastest moving particle must be taken for the calculation of the minimal UV fluence absorbed by the flowing fluid in the reactor, to make the most conservative estimate of the UV dose delivered to the syrup.

5.1.2.3 Residence time evaluation

The theoretical mean residence time was calculated as

$$\bar{t} = V_{\mathrm{reactor}}/Q \tag{5.1}$$

Table 5.6 UV reactor geometric parameters.

Parameter	in.	cm
Volume of 1 quartz sleeve (V_{sleeve})	89.04	1459.05
Cross-sectional area of quartz sleeve	1.58	10.21
Volume of the liquid ($V_{\mathrm{r}} - 7 \times V_{\mathrm{sleeve}}$)		14,719.37
Input pipe diameter	1.5	3.81
Wet cross-sectional area (A)	15.97	103.02
Hydraulic diameter (h)	1.09	2.78

Table 5.7 UV reactor flow conditions of liquid sucrose in the entrance pipe.

		$T = 80°F$			$T = 100°F$		
Flow rate (gpm)	Average velocity (m/s)	Density (m^3/kg)	Viscosity (Pa s)	Reynolds number	Density (m^3/kg)	Viscosity (Pa s)	Reynolds number
35.7	1.98	1350	0.187	540.7	1350	0.053	1918
20	1.11	1350	0.187	303	1350	0.053	1074
7.63	0.42	1350	0.187	115.2	1350	0.053	408
3.3	0.18	1350	0.187	50	1350	0.053	177

Table 5.8 Ultraviolet reactor flow conditions of liquid sucrose in the gap between UV lamps in the reactor.

		$T = 80°F$			$T = 100°F$		
Flow rate (gpm)	**Average velocity (m/s)**	**Density (m^3/kg)**	**Viscosity (Pa s)**	**Reynolds number**	**Density (m^3/kg)**	**Viscosity (Pa s)**	**Reynolds number**
35.7	0.22	1350	0.187	43.6	1350	0.053	155
20	0.12	1350	0.187	24.5	1350	0.053	87
7.63	0.05	1350	0.187	9.3	1350	0.053	33
3.3	0.02	1350	0.187	4.0	1350	0.053	14

For laminar flow, the minimum residence time in reactor can be evaluated from the mean residence time as $\bar{t}/2$. Also, minimum residence times for four selected flow rates were measured via pulse injections of a dye as a conservative tracer into the carrier fluid flow under steady-state flow conditions at the entrance of the reactor and measuring time from injecting dye until the colored liquid sucrose exits the reactor at the outlet. Technical limitations in the time of the experimental runs did not allow collecting samples to obtain RTD distribution. The measured minimum residence times were compared to calculated times and are given in Table 5.9.

The differences in theoretical and measured times indicate that actual minimum residence time was longer than the calculated residence time. The baffles installed in the center of the reactor for additional mixing can be a possible explanation for this effect. The measured residence times were used in further calculation of the UV dose delivered by the UV reactor in the liquid sucrose.

5.1.2.4 Spore recovery from sugar syrups

The recovery of spores from the sugar syrups was not affected by the holding time (2 h). *B. subtilis* spores were successfully recovered from sugar syrups using standard methods for enumeration described previously.

5.1.2.5 B. subtilis *spores inactivation*

B. subtilis spore inactivation data are summarized in Table 5.10. The data point summarizes the results of microbiological analysis of the samples taken from the tank after the run was complete. Thus these data represent the volume average inactivation of spores. For the purpose of process control, the samples were also taken directly from the pipe at the beginning, middle, and end of the processing runs. These samples were analyzed to verify that the level of inactivation was consistent through the process

Table 5.9 Residence time of liquid sucrose in the UV reactor at selected flow rates.

Flow rates	Theoretical residence times (s)		Measured residence times (s)	
gpm	Mean	Minimal	Mean	Minimal
35.7	6.5	3.3	NA	3.7 ± 0.57
20	11.7	5.8		
7.6	30.7	15.4		22 ± 1.0
3.3	70.7	35.4		39.33 ± 1.25

Table 5.10 Ultraviolet inactivation of *Bacillus subtilis* spores in liquid sucrose at 80°F with center lamp #7.

Flow rate (gpm)	Volume average $\log_{10}$ reduction	Time of the process run	Log_{10} reduction
3.3	3.01	Beginning	3.26
		Middle	2.98
		End	2.72
3.3	3.50	Beginning	3.62
		Middle	3.52
		End	3.70
3.3	2.80	Beginning	2.49
		Middle	2.67
		End	2.71
7.6	1.77	Beginning	1.72
		Middle	1.63
		End	1.75
7.6	1.54	Beginning	1.91
		Middle	1.55
		End	1.53
7.6	1.62	Beginning	1.66
		Middle	1.67
		End	1.62
35.7	0.55	Beginning	0.59
		Middle	0.53
		End	0.59
35.7	0.59	Beginning	0.63
		Middle	0.65
		End	0.61
35.7	0.50	Beginning	0.49
		Middle	0.46
		End	0.51

run. For purposes of calculating dose, the volume averaged inactivation data were used.

The inactivation of *B. subtilis* spores in liquid sucrose that include data in each test and average $\log_{10}$ reduction per pass for single-lamp, 7-lamp, and 3-lamp configurations are summarized in Tables 5.11 and 5.12.

The data in Table 5.11 showed that the mean spores' reduction larger than $5\log_{10}$ was achieved in the UV systems with all lamps turned on when sucrose was running at the maximum flow rate of 35.7 gpm. The further decrease in the flow rate resulted in zero counts of spores. In contrast, when one central lamp was left "ON," the maximum spores'

Table 5.11 Inactivation of *Bacillus subtilis* spores in liquid sucrose.

Number of UV lamps "ON"	Mean $\log_{10}$ reduction per pass at flow rates (gpm)					
	35.7		7.6		3.3	
1 Central lamp (#7)	0.55; 0.59; 0.5	0.55 ± 0.05	1.77; 1.54; 1.62	1.64 ± 0.12	3.01; 3.5; 2.8	3.1 ± 0.36
7 Lamps	5.56; 4.93; 5.24; 4.67	5.1 ± 0.39	>5	NA	NA	NA

Table 5.12 Inactivation of *Bacillus subtilis* spores in liquid sucrose in single-lamp and 3-lamp configurations.

UV reactor configuration, number of UV lamps "ON"	Mean $\log_{10}$ reduction per pass at flow rates (gpm)	
	7.6	3.3
1-Peripheral lamp (#4)	—	0.5
2-Side lamps (#2, 6, and 4)	3.7	—
3-Side lamps (#1, 3, and 5)	4.2	—

Table 5.13 Inactivation of *Bacillus subtilis* spores in HFCS.

Number of UV lamps "ON"	Mean $\log_{10}$ reduction per pass at flow rates (gpm)	
	15	1.5
Seven lamps	0.31	0.57

reduction of 3.1$\log_{10}$ was reached at the slowest flow rate of 3.3 gpm. Only slight differences in spores' inactivation were found (3.7 and 4.2 logs) when three side lamps were used in configurations 2 and 3 (Table 5.12) indicating relatively uniform exposure to UV light.

UV treatment of HFCS resulted in rather low inactivation of spores lower than 1 log at the slowest flow rate (Table 5.13) indicating the effect of the higher UV absorbance of HFCS in comparison with the UV absorbance of sucrose. After reviewing the data from individual pilot plant runs of HFCS at the flow rates of 1.5 and 15 gpm, a decision was made not to proceed with evaluating the UV treatment of HFCS due to the low efficacy of UV treatment in laminar flow regime.

5.1.3 Mathematical modeling of UV fluence rate

UV Calc, a software program for single and multiple-lamp UV reactors (courtesy of Bolton Photoscience Inc., Edmonton, AB, Canada), was used to compute three-dimensional fluence rate (irradiance) distribution, average fluence rate, and hence the fluence in the UV reactor. The program is based on the Multiple Point Source Summation Method with full accommodation of reflection and refraction at the air/quartz/water interface.

The calculations of UV fluence rate distribution and volume averaged fluence rate were made for the single central lamp reactor and for the multiple 7-lamp reactor. The input required was as follows:

- Transmittance of sucrose UV transmittance (UVT) = 35%;
- Lamp power of 320 W; efficiency of 35%;
- Lamp length of 157.5 cm;
- Lamp sleeve radius of 1.78 cm;
- Maximum cylinder radius of 7.34 cm;
- Refractive index of experimental medium of 1.372;
- Lamp sleeve medium; and
- Quartz with refractive index of 1.516.

5.1.3.1 Single-lamp UV reactor

The effect of percentage of transmittance of liquid sucrose in the range of 30%–40% UVT on UV fluence in the single-lamp reactor is shown in Fig. 5.4. The decrease in 10% of UVT resulted in a reduction of UV fluence of approximately 20%. The distribution of fluence rate (irradiance) in the radial distance in the UV reactor was also plotted and shown in Fig. 5.5 and shows the decrease in UV irradiance to almost 0 mW/cm^2 close to the walls of the chamber. The volume average fluence rate at 35% T was calculated equal to 3.61 mW/cm^{2}and was used for the UV fluence calculations. The longitudinal distribution of UV fluence rate was not significant.

5.1.3.2 Multiple-lamp UV reactor

UV fluence rate in the 7-lamp reactor was calculated using UV Calc 2 software. The input of each lamp coordinate was as follows (Table 5.14).

To calculate UV fluence in multiple-lamp reactors using UV Calc, the reactor was divided in four quadrants with the center at the origin. Since the quadrants were symmetrical, the UV fluence rate was computed for one quadrant. All other quadrants are equivalent. The data and chart are displayed only for the $+x$, $+y$ of one quadrant. The fluence (UV dose)

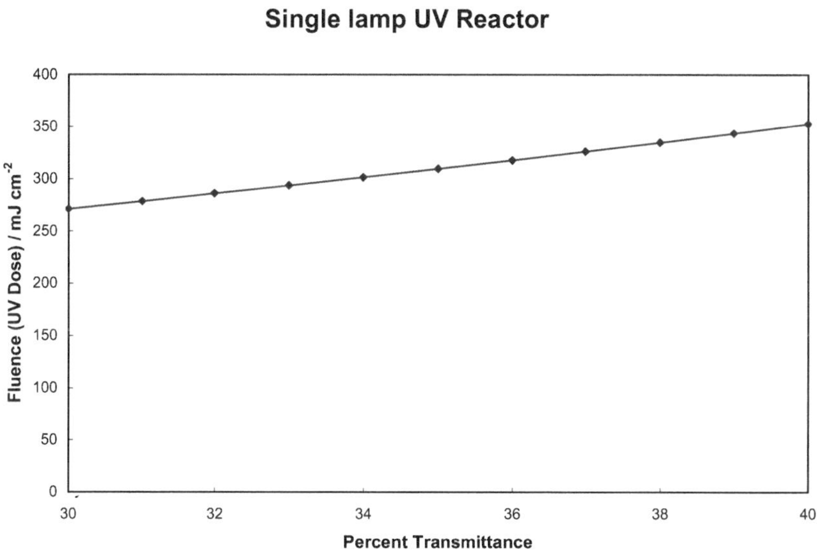

Figure 5.4 Effect of liquid sucrose transmittance on UV fluence in the single-lamp reactor.

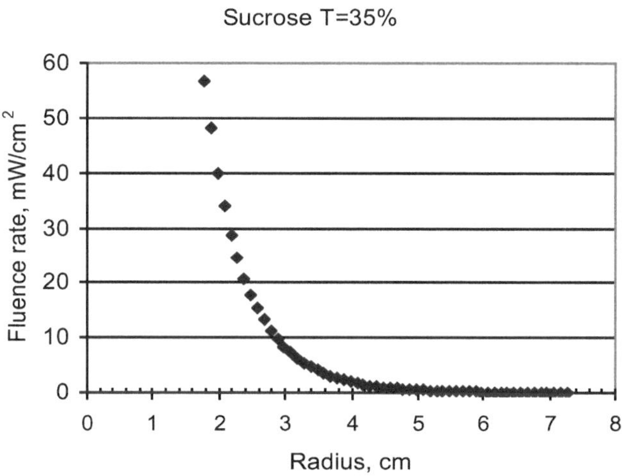

Figure 5.5 UV fluence radial in liquid sucrose in the configuration with the single central lamp.

estimate is the maximum theoretical value, since no account was taken of mixing inefficiencies. Given the high viscosity of the sucrose, the mixing efficiency will be quite low.

Table 5.14 Ultraviolet lamps coordinates in the 7-lamp reactor configuration.

Lamp number	1	2	3	4	5	6	7
X coordinate (cm)	2.54	5.08	2.54	−2.54	−5.08	−2.54	0
Y coordinate (cm)	4.39	0	−4.39	−4.39	0	4.39	0

Table 5.15 Volume averaged fluence rate in the 7-lamp reactor.

Fluence rate distribution values are listed in units of	mW/cm^2	
Fluence rate distribution value plotted	39.8	%T
Experimental medium: water → refractive index	1.372	
Lamp sleeve medium: quartz → refractive index	1.516	
Number of lamps for this calculation	7	
Longitudinal correction factor	1.000000	
Average fluence rate (uncorrected)	37.7	mW/cm^2
Average fluence rate (corrected)	37.75	mW/cm^2
Flow rate, *Q*	35.7	gpm
	135.15	L/min
	2.25	L/s
Reactor volume, *V*	14.96	L
Hydraulic residence time, *t*	6.65	s
Fluence (UV dose)	250.7	mJ/cm^2

The results of the calculation are given in Table 5.15. The volume average fluence rate for the 7-lamp reactor was found to be equal to 37.7 mW/cm^2. A 3-D plot showing distribution of the fluence rate for the 7-lamp reactor case is given in Fig. 5.6.

To compute the volume average fluence in the UV reactor, the following technical parameters and factors that are summarized in Table 5.15 were used.

The volume average theoretical UV fluence (UV dose) delivered in 7-lamp UV reactor was calculated as a product of average fluence rate and hydraulic residence time and was equal to 250.7 mJ/cm^2.

5.1.3.3 Decimal reduction UV dose evaluation

Decimal reduction dose (D_{10}) of *B. subtilis* spores (ATCC 6633) in liquid sucrose or the effective germicidal dose required for 90% inactivation was determined based on biodosimetry approach for the single-lamp source configuration.

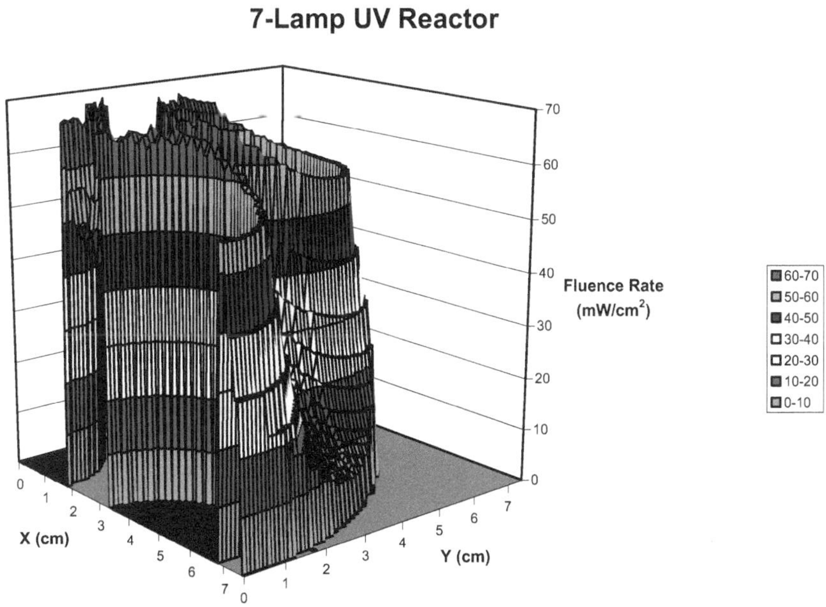

Figure 5.6 Fluence rate distribution in the UV reactor with seven lamps.

5.1.3.4 Mathematical calculations

The calculations were performed for three flow rates and corresponding residence times and knowing mean microbial log reduction by the procedure summarized in Table 5.16.

1. The mean residence time $t = V_{\text{reactor}}/Q$ was calculated and measured for flow rates of 3.3, 7.6, and 35.7 gpm. Since the flow was in laminar regime, the flow pattern was assumed to be parabolic.
2. The distribution of UV fluence rate in the single-lamp reactor was used from calculation by UV Calc 2 (Bolton Photoscience Inc, Edmonton, AB, Canada):
   a. Volume averaged fluence rate: 3.61089 mW/cm^2.
3. The volume averaged fluence rate (irradiance) was multiplied by the hydraulic residence times to estimate the average fluence delivered in the reactor.
4. The number of log reductions ($\log_{10}N_0 - \log_{10}N$) of *B. subtilis* spores per one pass was obtained experimentally for each flow rate (Table 5.11). The dose–response plot in Fig. 5.7 shows $\log_{10}$ survival plot at the tested flow rates.

Table 5.16 Data used for the D_{10} evaluation based on the volume average fluence.

Flow rate (gpm)	Mean residence time (s)	Volume average UV fluence rate (mW/cm^2)	Volume average UV fluence (mJ/cm^2)	Mean log reduction per 1 pass
3.3	70.7	3.61	255.2	3.01
3.3	70.7	3.61	255.2	3.5
3.3	70.7	3.61	255.2	2.8
7.6	30.7	3.61	110.8	1.77
7.6	30.7	3.61	110.8	1.54
7.6	30.7	3.61	110.8	1.62
35.7	6.5	3.61	23.5	0.55
35.7	6.5	3.61	23.5	0.59
35.7	6.5	3.61	23.5	0.5

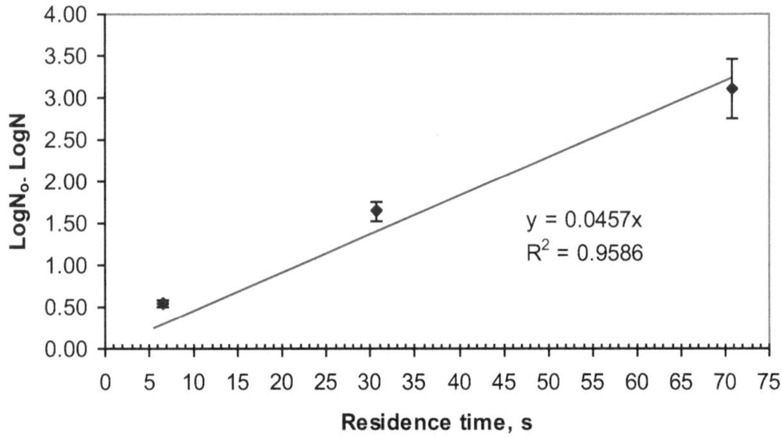

Figure 5.7 Survival of *Bacillus subtilis* spores (mean $\log_{10}$ reduction vs time) in the UV reactor with a single central lamp.

The linear character of the survival plot for *B. subtilis* spores demonstrated that the first-order inactivation model can be used to calculate inactivation rate constant based on the following Eqs. (5.2) and (5.3):

$$\frac{dN}{dt} = -kIN \tag{5.2}$$

By integration,

$$\frac{N}{N_0} = \exp(-kIt) \tag{5.3}$$

$N = N_{av}$ was measured experimentally; parameters in the inactivation kinetics model were obtained by mathematical methods, such as the least square method. The purpose of least squares method is to determine the inactivation constant in the equation that minimize the sum of the squares of the deviations between logarithm reduction values fitted from the equation and the experimental values:

$$SD = \sum \left[\log\left(\frac{N_{av}}{N_0}\right)_{fit} - \log\left(\frac{N_{av}}{N_0}\right)_{exp} \right]^2 \tag{5.4}$$

Because only one constant needed to be fitted, trial and error method is used. At first, the maximum and minimum inactivation constants are proposed according to experience and form an interval $[k_{min}, k_{max}]$. Next, the interval $[k_{min}, k_{max}]$ is divided into 100 individual points evenly spaced, k_i ($i = 1, 2, \ldots, 100$). Then, a point with the least standard deviation can be obtained and a new interval can be formed by extending two points forward and backward from the point with the least standard deviation. If the point with the least standard deviation lies at or near boundary points, it shows that the best inactivation constant perhaps lies beyond the current interval and broader interval is needed. Above steps are repeated until the required accuracy is met. The source code using C language was used to calculate inactivation rates.

After fitting experimental data with the first-order inactivation model, a value for the rate constant $k = 0.029150$ cm^2/mJ with the coefficient of determination $R^2 = 0.931$ was obtained. The maximum absolute error of $\log_{10}$ reduction is 1.19 when the $\log_{10}$ reduction is 4.9. The minimum relative error of $\log_{10}$ reduction was 7.33% when log reduction is 3.01.

According to the calculations, the D_{10} can be found as

$$D_{10} = 2.3/k = 78.9 \ \text{mJ/cm}^2 \tag{5.5}$$

5.1.3.5 Biodosimetry approach of UV dose evaluation in the multiple-lamp UV reactor

Knowledge of the D_{10} value and mean $\log_{10}$ reduction of bacterial spores in liquid sucrose when seven lamps were "ON" (Table 5.11) allowed for the calculation of the volume averaged absorbed dose (UV dose) delivered by the UV reactor as

$$\text{UV dose} = D_{10} \times \log(N/N_0) = 5.1 \ \text{logs} \times 78.9 \ \text{mJ/cm}^2 = 402.4 \ \text{mJ/cm}^2 \tag{5.6}$$

The summary of UV dose validation for the inactivation of *B. subtilis* spores in liquid sucrose using biodosimetry and mathematical modeling approaches is presented in Table 5.17.

The observed differences in the resulted values of the fluence delivered in UV reactor obtained through biodosimetry and theoretical evaluation may result from the following assumptions:

1. 35% UVT at 253.7 nm of sucrose was used in calculations;
2. Nonuniform fluence rate distribution in the UV reactor with a single central lamp;
3. Nonuniform RTD due to laminar flow of sucrose in the UV chamber;
4. Product UVT and viscosity are critical product parameters of the treatment;
5. Process flow rate, number of UV sources, their geometrical distribution in the chamber, and lamp intensity are critical process parameters to consider; and
6. Overestimated D_{10}—value of *B. subtilis* spores in sucrose that was calculated using data from experimental runs in a large reactor where nonuniformity of UV treatment is high.

5.1.3.6 Numerical simulation of distribution of sucrose velocity and spores' survivors in a 7-lamp UV reactor configuration

Due to nonuniform distribution of velocity in laminar flow and UV fluence rate distribution in outlet of the tested reactor, the numerical simulation of the sucrose velocity and spores survivors distribution using Fluent 6.2 software was conducted. The results of the calculation of velocity and the distribution of survivors in the 1/6 of the reactor volume due to reactor symmetry are shown in Figs. 5.8 and 5.9 at the highest flow rate of 35.7 gpm.

Table 5.17 Fluence delivered in liquid sucrose in 7-lamp UV system.

Approach	Calculations	Fluence (mJ/cm^2)
Mathematical modeling	UV Calc Fluence rate (mW/cm^2) × residence time (s)	250.7
Biodosimetry	I-log reduction dose (mJ/cm2) × microbial log reduction	402.4

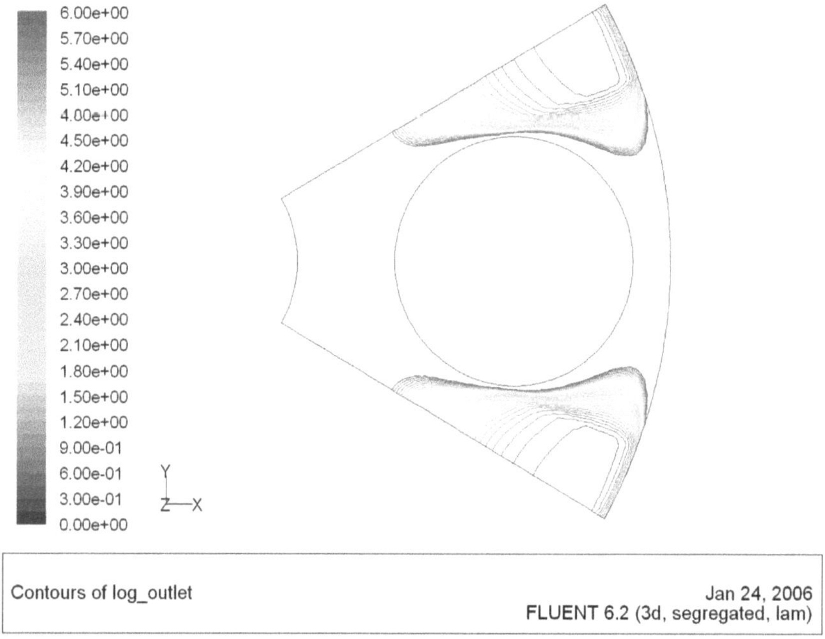

Figure 5.8 Velocity distribution of sucrose at 35.7 gpm in 1/6 of the reactor volume.

The nonuniformity of UV treatment and spores inactivation can be seen in the velocity distribution resulted in a large variation of survivors across the UV reactor. As expected, the highest values of inactivation were observed in the close distance to UV light sources. It was found that the point with the most surviving spores is at coordinates of $x = 5.96$ cm and $y = 3.37$ cm where $\log_{10}$ reduction equals 0.2. At this point, axial velocity is 0.349 m/s, 1.6 times as much as average velocity (0.219 m/s). Actually, the velocity field was not the main factor, which caused broad distribution of $\log_{10}$ reduction at the outlet. The fastest particle velocity was equal to 0.418 m/s, about two times as much as average velocity (see Fig. 5.8). The more important factor causing broad distribution of $\log_{10}$ reduction was effects of the absorption coefficient of sucrose. The lowest fluence rate at the point with most surviving spores was 0.0072 mW/cm^2.

5.1.4 Conclusions

Microbial challenge studies of inactivation of *B. subtilis* in the solution of liquid sucrose showed the feasibility of UV treatment at 253.7 nm in laminar flow to reduce risk of product contamination by thermal resistant

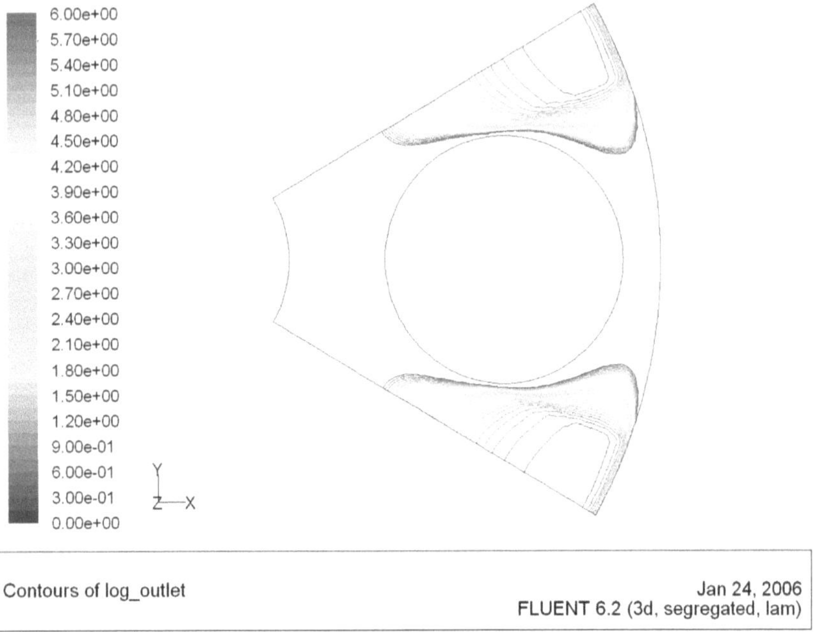

Figure 5.9 Distribution of $\log_{10}$ of *Bacillus subtilis* spores survivors in the outlet of the ultraviolet reactor shown in 1/6 reactor volume.

spores. The theoretical and biodosimetry evaluation of UV dose resulted in the value between 250 and 400 mJ/cm^2 being sufficient to achieve 5log reduction of spores. The mathematical modeling of residence time and spore survivors' distribution demonstrated the nonuniformity due to optical properties and high viscosity of sucrose syrup that are main critical product parameters of UV treatment.

5.2 UV dose verification studies in commercial-scale system

5.2.1 Background

In the next step, a bench top singular lamp annular UV reactor was designed and built (American Ultraviolet Company) equipped with a low-pressure amalgam arc lamp at 253.7 nm that was similar to the lamps used in the multiple-lamp commercial reactor to determine inactivation performance and the UV dose delivered to liquid sucrose using *B. subtilis* spores as the test microorganism and more accurately evaluate UV dose for 1log reduction of these spores in liquid sucrose. In the biodosimetry tests, the sucrose syrup inoculated with *B. subtilis* spores was pumped through the bench reactor at various flow rates, and spore counts were determined before and after UV treatments. Based on the calculated average UV fluence and fluence rate in the reactor annulus and microbial reduction data, the dose–response curve was constructed and a decimal reduction dose or D_{10} value was determined. The obtained D_{10} value was used to verify the UV dose delivered in the multiple-lamp pilot-scale reactor.

The objectives of the verification study were as follows:

1. To develop and build a singular lamp laboratory-scale UV reactor to scale down multiple-lamp commercial UV system and conduct biodosimetry study in liquid sucrose;
2. To measure inactivation kinetics of *B. subtilis* spores and to determine its UV D_{10} using a singular lamp annular bench-scale reactor; and
3. To determine dose delivered by the bench-scale UV reactor and to verify UV dose delivered in the multiple-lamp pilot-scale reactor.

5.2.2 Materials and methodologies

5.2.2.1 UV processing equipment

The experimental setup included the following components and was equipped with measuring instruments that are schematically shown in Fig. 5.10.

1. Annular UV reactor (American Ultraviolet, Lebanon, MI),
2. Power supply with transformer/ballasts (American Ultraviolet, Lebanon, MI),
3. Watson-Marlow Bredel series 704 IP55 Peristaltic Pump (Watson-Marlow, Wilmington, MA),
4. Digital flow meter (Great Plains Industries, Wichita, KS),
5. Inlet Tank—Nalgene Carbuoy with sanitary outlet connection (Nalgene, Rochester, NY),
6. 1″ Bore, 4.8 mm wall #92 Marprene Tubing (Watson-Marlow, Wilmington, MA),
7. Two stainless-steel hose barb/sanitary adapters 1″ tri-clamp × 1″ Barb (Holland Applied Technologies, Burr Ridge, IL),
8. Steel hose clamps (generic; Home Depot, Ford City, IL),
9. Stainless-steel piping with sanitary fittings and tri-clover clamps (unknown, from the pilot plant), and
10. Crescent wrench (generic).

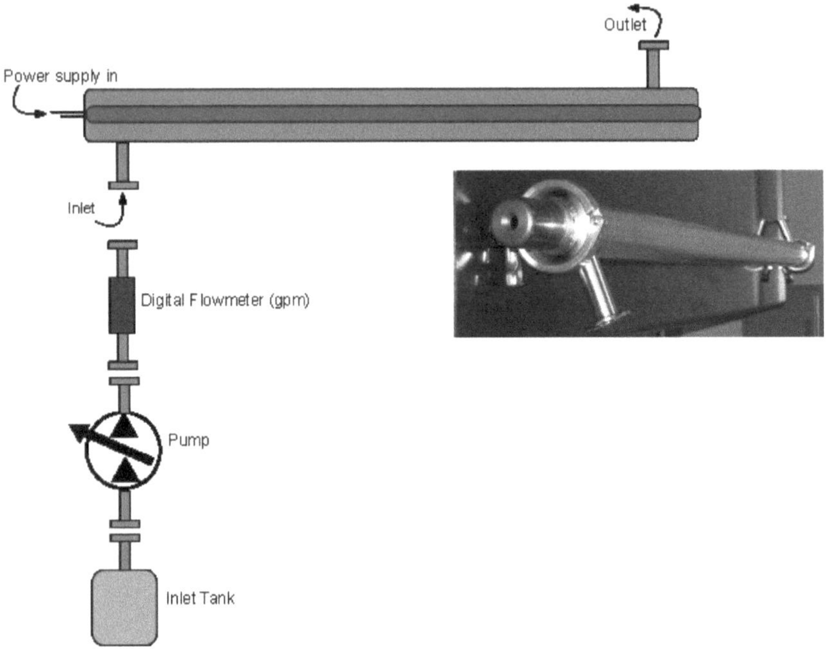

Figure 5.10 Bench-scale UV reactor experimental setup.

Table 5.18 UV reactor technical and geometrical characteristics.

Parameter	in.	cm
Number of chambers	1	
Peak flow rate	5 gpm	
Number of lamps	1	
Outer diameter of the reactor	2.52″	6.4 cm
Working length (*L*)	57.25″	145 cm
Full length	66.5″	168.9 cm
Inner diameter of the reactor	2.21″	5.61 cm
Total cross-sectional area of the reactor		24.68 cm^2
Cross-section of the annulus		14.47 cm^2
Total volume of UV reactor V_r		2067 cm^3
Outer diameter of the quartz sleeve	1.417″	3.6 cm

5.2.2.2 Specifications of UV reactor

The technical characteristics and dimensional parameters of the singular lamp annular UV reactor (American Ultraviolet Co., Lebanon, MI) are given in Table 5.18.

5.2.2.3 UV lamp

The lamp was a 320-W output power low-pressure, high-intensity amalgam lamp. According to the manufacturer, 90% of UV output is within the range 233.7–273.7 nm. The UV light output of the lamp is approximately 105–110 W and produces 750 $\mu W/cm^2$ at a distance of 1 m in the air, which is a 35% efficiency rating. The lamp warranty is 9000 h. The lamp arc length is 149 cm (58.66″) with the OD of 19 mm. The lamp sleeve is type 214 clear fused quartz circular tubing rated for the transmission of 94%.

5.2.2.4 Measurements of UV fluence rate on the surface of UV lamps

After the UV lamp was turned on for 10 min, an IL 1700 Research Radiometer equipped with an SUD240 detector (International Light, Inc., Newburyport, MA) was used to measure the incident irradiation fluence rate of the lamp. The fluence rate was measured along the entire length of the quartz sleeve and averaged.

5.2.2.5 Pump flow rate calibration

The flow rate was controlled by and read from the digital flow meter attached to the reactor inlet. The pump flow rates were measured using sucrose at 20°C at a given pump speed (rpm) before proceeding with inoculation experiments. A "bucket and stopwatch" method was used in which flow rate is calculated by measuring actual sucrose volume pumped during a measured period of time. Each measurement was made 3–5 times, depending on the flow rate. The plot of the pump calibration in the range from 50 to 230 rpm is given in Fig. 5.11. The results demonstrated that repeatable and reliable flow could be obtained in the range from 50 to 140 rpm or 1.34 to 5.6 gpm.

5.2.2.6 Test product

Liquid sucrose was received from Savannah Foods & Industry (Port Wentworth, GA). Products were received in 55-gallon drums with a net weight of 625 lbs. Lot #S343F was used for laboratory-scale experiments.

5.2.2.7 Optical and physical product parameters

Absorbance of sucrose syrup was measured at 254 nm using double-beam spectrophotometers (Cary 100 by Varian, Walnut Creek, CA; Lambda 2 by Perkin Elmer, Boston, MA). Due to the low light transmittance, sample solutions were evaluated using matched demountable fused quartz

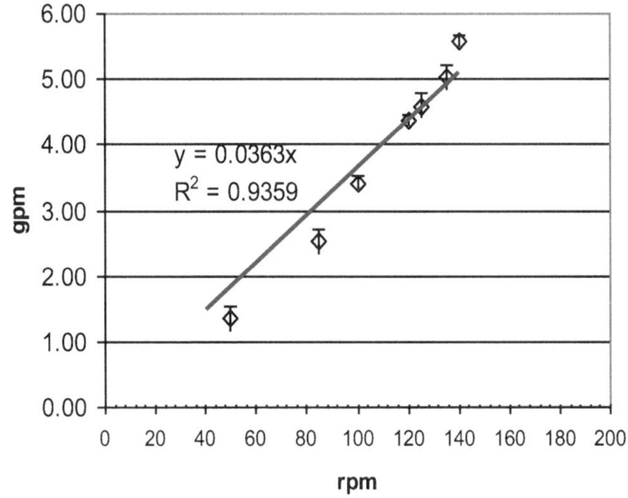

Figure 5.11 Calibration plot of the peristaltic pump rpm versus gpm units.

cuvettes (NSG Precision Cells Inc., NJ) with path lengths of 0.1, 0.2, 0.5, and 1.0 mm. The absorption coefficient of the sample solution was determined from the slope of the absorbance versus path length.

5.2.2.8 Test microorganisms

B. subtilis ATCC 6633 spores retained from stage 1 of this study were used for the construction of the dose–response curve. This microorganism was selected because the UV inactivation response of *B. subtilis* spores in water has been well documented and it was used in the pilot-scale studies of the multiple-lamp UV reactor in the stage 1 trials.

5.2.2.9 Inoculum preparation

5.2.2.9.1 Bacillus subtilis

B. subtilis ATCC 6633 spore crops prepared for stage 1 of this study were also used in stage 2 experiments (4).

5.2.2.9.2 Product inoculation

Sucrose syrup was added to the inlet tank (Fig. 5.10). For each test run, the test inoculum was added to c. 1000 mL of sucrose syrup in a sterile container and mixed for 2–3 min by stirring with a sterile pipette. This inoculated sucrose syrup mixture was then added to the sucrose syrup in the inlet tank to obtain a target inoculum level of 10^5 CFU/mL for experiments conducted with the bacteria and yeast cocktails, and

10^4 CFU/mL for experiments conducted with mold spores. Inoculated syrup was mixed in the covered mixing tank for a minimum of 5 min to ensure even distribution of inoculum in the syrup.

5.2.2.10 Processing parameters

To scale down the flow regime of the 7-lamp pilot-scale UV reactor for the bench-scale treatment experiments, the flow rates were calculated assuming that the Re numbers should be similar in both commercial and bench reactors. Knowing the Re numbers in the multiple-lamp reactor (ranging from 185 to 18), the mean velocity of the liquid sugar in the singular lamp reactor was calculated first using the formula: $\bar{v} = \frac{\mathrm{Re}\mu}{\rho d}$, where sucrose dynamic viscosity $\mu = 0.19$ Pa s and density $\rho = 1350\ \mathrm{kg/m^3}$ and $d = 0.01$ m is the characteristic dimension or annulus gap of the reactor.

The volumetric flow rate was calculated as $\dot{V} = \bar{v} F_{\mathrm{annulus}}$, where F_{annulus} is a cross-section area of the annulus. The obtained flow rates range was between 5.0 and 0.4 gpm. The flow rates, mean residence time in the reactor, and sample volume used as variables in the inoculation experiments are given in Table 5.19. Product temperature for all experiments was 80°F.

5.2.2.11 Sampling

Triplicate samples of inoculated, untreated sucrose syrup were aseptically removed from the inlet tank and transferred to sterile sample containers for enumeration to determine the initial levels. For the first 45 s, product through the UV reactor was collected in the outlet tank. After this initial run time, the sucrose syrup was diverted to a sterile carboy. Volume averaged treated samples were aseptically removed from the carboy for microbial

Table 5.19 Experimental variables: flow rates and residence time.

Rpm	Flow rate (gpm)	Flow rate (mL/s)	Residence time (s)	Volume of sample (gallons)
140	4.9	308.7	6.7	7
125	4.3	270.9	7.63	6
100	3.5	220.5	9.37	5
85	2.4	151.2	13.6	4
35	1.0	63.0	32.8	3
20	0.5	31.5	65.6	2

analysis. Due to the limited volume of *B. subtilis* spore crop remaining after phase 1 of this study, each spore run was conducted in duplicate.

5.2.2.12 Enumeration procedures

Bacterial test microorganisms were quantified by preparing serial dilutions in PW, spread plating onto TSA plates, and incubating at 35°C for 48 h. Yeast and mold were quantified by preparing serial dilutions in PW, spread plating onto acidified potato dextrose agar (PDA), and incubating at 20°C–25°C for 5 days.

5.2.3 Results

The average irradiation fluence rate measured on the surface of the quartz sleeves was 55 mW/cm^2. UV output as reported by the manufacturer was approximately 112 W. According to this output, the calculated average irradiation fluence rate should be 35 mW/cm^2 (112 W/3241 cm^2), which is lower than the measured value. However, the manufacturer tested output at 1 m away from the lamp only and guaranteed it was more than 750 μW/cm^2. According to the manufacturer recommendations, the intensity factor is 77.2 (1). Based on these considerations, estimated average incident fluence rate on the surface of the quartz sleeve is approximately 58 mW/cm^2. The measured values of fluence rate on the tube surface were used in the following calculations.

5.2.3.1 Sucrose absorbance

The variations of the sucrose absorption coefficient (α) were evaluated by taking samples from three lots #S 343F; S 159F, and S 305F. The absorption coefficient of the sucrose from the lot #S 159F (used in the pilot-scale studies) was equal to 0.421/cm. The absorption coefficient of $\alpha = 0.560$/cm was obtained in the lot #S 343, used in the laboratory-scale studies. The measured values from two lots indicated that the variations of 12%–15% can be noted for the sucrose absorption coefficient. Fig. 5.12 shows the absorption spectrum of sucrose and fructose for the wavelength range from 150 to 500 nm. It can be seen that the maximum of UV light absorbance of sucrose is located at approximately 200 nm and decreases with increase of wavelength.

The absorbance values of fructose in 1 cm cuvette were much higher than for sucrose. For this reason, 1-mm cuvette was used. The maximum absorption was at approximately 280 nm with a minimum at about 250 m.

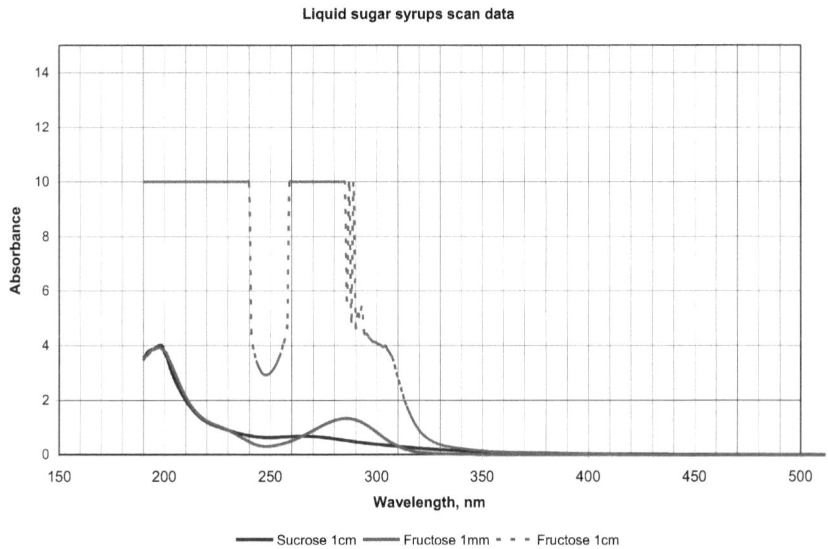

Figure 5.12 The absorption spectrum of sucrose syrup and fructose in the wavelength range from 150 to 500 nm.

5.2.3.2 UV fluence determination

UV fluence H can be determined as a product of UV fluence rate I, mW/cm^2, in the sucrose and mean residence time t [Eq. (5.7)],

$$H = I \times t \ (mJ/cm^2) \tag{5.7}$$

The mean residence times in the UV reactor can be determined as: $t = V/Q$, where V is the volume of the reactor annulus (cm^3) and Q is the sucrose flow rate (mL/s).

The Beer–Lambert equation can be used to calculate the UV fluence rate at the radial points along the radius of the annular UV reactor [Eq. (5.8)]:

$$I_r = I_0 \frac{R_1}{r} \exp[-\alpha(r - R_1)] \tag{5.8}$$

where I_r is UV fluence rate at a distance r from center (mW/cm^2); I_0 is UV fluence rate at the surface of the quartz sleeve (mW/cm^2); $I_0 = 55\ mW/cm^2$; r is radial distance between the center of the source to the point at which irradiance is measured, $R_2 = 2.805$ cm—internal radius of the UV chamber; R_1 is the radius of the central UV lamp quartz sleeve, $R_1 = 1.8$ cm; 1.8 cm $< r <$ 2.8 cm. Penetration depth is equal to

$R_2 - R_1 = 1$ cm; α is the absorption coefficient of the sucrose (1/cm); $\alpha = 0.421$/cm.

Velocity profile for laminar flow can be obtained from Eq. (5.9) (Bird et al., 2002).

$$u(r) = C_1\left[1 - \frac{r^2}{R_2^2} + \frac{1-\kappa^2}{\ln(1/\kappa)}\ln\left(\frac{r}{R_2}\right)\right]U_{av} \tag{5.9}$$

where U_{av} is the average velocity, $U_{av} = \frac{Q}{\pi(R_2^2 - R_1^2)}$ (cm/s); C_1 is the constant determined by κ, $C_1 = \frac{2}{\frac{1-\kappa^4}{1-\kappa^2} - \frac{1-\kappa^2}{\ln(1/\kappa)}}$; κ, ratio of the radius of the inner cylinder to that of outer cylinder, $\kappa = R_1/R_2$.

The plots of UV fluence rate, velocity, residence time, and UV fluence distribution along the radius in the single-lamp UV chamber are given in Fig. 5.13.

Next, the average UV fluence rate (I_{av}) within the annular reactor was estimated using Eq. (5.10). Fluence rate in each liquid element (I_r) with $r = 0.01$ cm was multiplied by the volume of that element (V_r) and

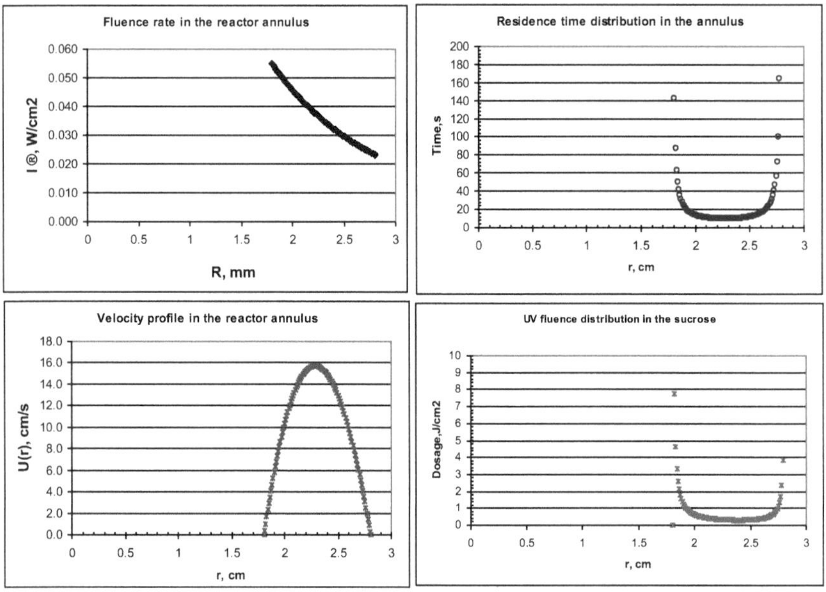

Figure 5.13 Distribution of UV fluence rate, velocity, residence time, and UV fluence in the liquid sucrose in UV reactor annulus.

summed over each element of solution and divided by total volume of the reactor (V).

$$I_{av} = \frac{\sum_{R_1}^{R_2} I_r V_r}{V} \tag{5.10}$$

From this formula the obtained magnitude of volume averaged fluence rate in the sucrose in the annulus gap was equal to $I_{av} = 27.41\ \text{mW/cm}^2$. This value was used in the further calculations of the volume average fluence using Eq. (5.7) for the flow rates tested. It should be noted that the estimate of the obtained fluence (UV dose) is the maximum theoretical value, since no account is taken of mixing inefficiencies.

5.2.3.3 *UV inactivation kinetics of* B. subtilis *spores*

Studies were conducted to determine the level of inactivation of *B. subtilis* spores after one pass through the annular reactor for each tested flow rate. Flow rates from 2.4 to 4.9 gpm (151.2–308.7 mL/s) were used to construct a dose–response curve. Knowledge of UV fluence at tested flow rates in the annulus allowed the calculation of the D_{10} of *B. subtilis* spores or effective germicidal dose required for 90% destruction. The D_{10} value of *B. subtilis* spores was calculated from the slope of the constructed dose–response curve (Fig. 5.14) that described the inactivation of *B. subtilis* spores as a function of absorbed UV dose. Table 5.20 summarizes the results. For the selected flow rates, each test was run in duplicate.

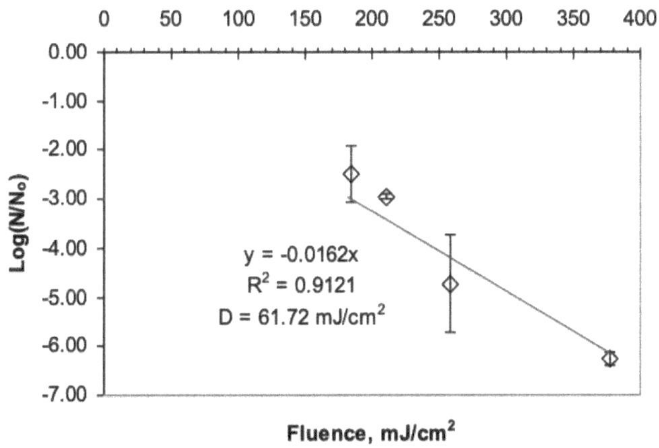

Figure 5.14 Dose–response curve of *Bacillus subtilis* spores survival in liquid sucrose treated in the annular UV reactor.

Table 5.20 UV inactivation of *Bacillus subtilis* spores in the annulus reactor.

Flow rate (mL/s)	Residence time (s)	UV fluence rate (mW/cm^2)	UV fluence (mJ/cm^2)	Log_{10} reduction per 1 pass, 1 and 2 runs		Average log_{10} reduction	Standard deviation
308.7	6.70	27.41	183.5	2.28	2.69	2.48	0.29
270.9	7.63	27.41	209.1	2.93	2.97	2.95	0.03
220.5	9.37	27.41	256.9	5.08	4.39	4.74	0.49
151.2	13.67	27.41	374.7	6.21	6.30	6.25	0.07

5.2.3.4 UV dose delivered by the pilot-scale 7-lamp UV reactor

Knowledge of UV sensitivity of *B. subtilis spores* to UV light in sucrose expressed as D_{10} value of 61.7 mJ/cm^2 allowed re-calculating and verifying the UV dose delivered by the pilot-scale 7-lamp UV reactor. The delivered or effective UV dose is a product of D_{10} and logarithmic count reduction of spores $\log(N/N_0)$. According to the data obtained in phase 1 of the project, at the maximum flow rate of 35.7 gpm when all seven lamps of the systems were "ON," the obtained mean $\log_{10}$ reduction of bacterial spores was equal to 5.1 logs. This means that effective UV dose in the commercial UV system is

$$\text{UV dose} = D_{10} \times \log\left(N/N_0\right) = 5.1 \ \text{logs} \times 61.7 \ \text{mJ/cm}^2 = 314.8 \ \text{mJ/cm}^2 \tag{5.11}$$

The obtained value of the delivered UV dose based on the biodosimetry data using laboratory-scale single-lamp reactor is lower and closer to the theoretical value of 250.7 mJ/cm^2 than the UV dose reported in the phase 1 due to mixing inefficiencies and nonuniformity issue in the multiple-lamp reactor (Table 5.21). Using single-lamp continuous system with similar flow and lamp characteristics can be recommended as a biodosimetry approach for low UVT food liquids.

5.2.3.5 Modeling of UV inactivation kinetics of B. subtilis spores

Various modeling approaches have been proposed to describe and predict UV inactivation kinetics. Among them, the first-order inactivation model is the simplest. The inactivation rate constant of *B. subtilis* was determined by fitting experimental data using a first-order and series-event reaction models (Fig. 5.15).

Table 5.21 Comparison of the values UV fluence delivered in 7-lamp UV system.

Approach	Calculations	Fluence (mJ/cm^2)	% difference
Mathematical modeling	UV Calc Fluence rate (mW/cm^2) × residence time (s)	250.7	
Biodosimetry in multiple-lamp system	I-log reduction dose (78.9 mJ/cm^2) × microbial log reduction	402.4	36
Biodosimetry in single-lamp system	I-log reduction dose (61.7 mJ/cm^2) × microbial log reduction	314.8	20.3

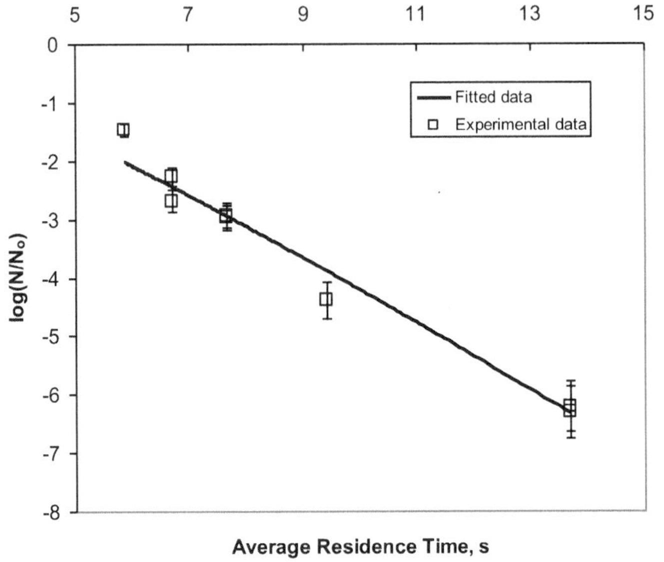

Figure 5.15 Inactivation curve of *Bacillus subtilis* spores in the annular UV reactor.

5.2.3.6 First-order inactivation model

This model assumes that the inactivation rate changes with respect to pathogen concentration, N, and fluence rate, I, such that

$$\frac{dN}{dt} = -k_1 IN \tag{5.12}$$

where k_1 is the first-order inactivation constant (cm^2/mJ). The first-order inactivation reaction was also defined as pseudo-first-order model or mixed second order model (Severin et al., 1983). If k_1 and I are constant, by integration,

$$\frac{N}{N_0} = \exp(-k_1 It) \tag{5.13}$$

The first-order model was able to reasonably predict microbial inactivation when the fluence was within certain limits. Sigmoidal shaped inactivation curves were often observed and reported. At first, the value of the first-order inactivation constant is low at relatively low UV fluence. The first-order inactivation constant increased with fluence and remained constant within a certain fluence range. Finally, when the fluence was larger than a certain value, the first-order inactivation constant decreased with the increase in UV fluence. This phenomenon is referred to as tailing phenomena.

A lag in microbial inactivation at low levels of UV fluence (the shouldered survival curve) can be observed because microorganisms exposed to a sublethal UV fluence may repair injuries and continue multiplying. The sublethal UV fluence has a slight adverse effect with the analytical procedure to quantify microbial viability. Also, the tailing phenomena (the decline in the slope of the fluence-log reduction curve) at high UV fluence may be observed. The tailing phenomena may be attributable to the result of heterogeneity among a population of microorganisms. Some organisms may be relatively less resistant to UV radiation while other organisms in the same population may be more resistant. Another reason for the tailing effect is presence of particles such as pulp in fresh juices. Particles may serve as a hiding space for a viable organism or an opaque surface to shade microorganisms from UV radiation.

Because the tailing phenomenon is mostly observed when the log reductions are larger than 5, it is not very important in commercial applications of UV disinfection. Other models, for example, series event (Severin et al., 1983) was developed to account for deviations from the first-order model at low UV fluence.

5.2.3.7 Series-event inactivation model

As stated before, a lag in microbial inactivation at low UV fluence can be observed for various kinds of microorganisms. The series-event inactivation model was proposed by Severin et al. (1983) to account for the lag at low fluence. It assumes that inactivation of microorganism elements takes place in a stepwise fashion,

$$M_0 \xrightarrow{k_{SE}I} M_1 \xrightarrow{k_{SE}I} \cdots M_i \xrightarrow{k_{SE}I} \cdots M_{n-1} \xrightarrow{k_{SE}I} M_n \xrightarrow{k_{SE}I} \cdots \tag{5.14}$$

The inactivation rate at each step is the first order with respect to the fluence rate I,

$$\frac{dN_i}{dt} = k_{SE}I(N_{i-1} - N_i) \tag{5.15}$$

where subscript i is event level and k_{SE} is the inactivation constant in the series-event inactivation model. k_{SE} is assumed to be the same for different event levels. When n elements (a threshold) of microorganisms have been inactivated, the microorganisms will become nonviable. If k_{SE} and I are constant, the concentration of surviving microorganisms N is determined by Eq. (5.16):

$$\frac{N}{N_0} = \exp(-k_{SE}It)\sum_{i=0}^{n-1}\frac{(k_{SE}It)^i}{i!} \tag{5.16}$$

where n is a threshold. It is obvious that if $n = 1$, the above equation will be reduced to the first-order model.

The physical meaning behind Eq. (5.16) is that more than one hit is required for UV inactivation of an individual microorganism. At the beginning of UV inactivation process (low fluence), the probability of individual microorganism to obtain n hits (where $n > 1$) is rather low. As the UV inactivation continues, more surviving microorganisms have accumulated $n - 1$ hits and require only one additional hit to be inactivated completely (Harm, 1980) so that the inactivation curve becomes steeper with increasing fluence. Larger threshold values of n represent microorganisms that are more resistant at low UV fluence.

The shouldered survival curve can be observed in the UV inactivation of many types of microorganisms. The first-order model ($n = 1$) was valid only for some viruses whose sensitive material is single-stranded DNA or single-stranded RNA (Harm, 1980). For example, it was reported that $n = 1$ for f2 bacterial virus (Severin et al., 1984). Both the first-order inactivation model and series-event inactivation model (Severin et al., 1983) were used in numerical simulations of inactivation rate constant of spores.

The results of computation of inactivation rate constants are summarized in Table 5.22. When eight experimental points of *B. subtilis* inactivation data were fitted by the series-event inactivation model with threshold $n = 4$, the obtained value of the inactivation constant was $k_{SE} = 0.094$ cm^2/mJ with $R^2 = 0.97$. The resulted value of the decimal reduction dose or D_{10} of *B. subtilis* was in the range of 23.86–24.47 mJ/cm^2. The UV dose for a one log reduction of *B. subtilis* ATCC 6633 reported by Chang et al. (1985)

Table 5.22 Inactivation parameters of *Bacillus subtilis* spores.

Event level, n	Inactivation rate k (cm^2/mJ)	Global min SSD	R^2	D_{10} (mJ/cm^2)
1	0.0204	75.30	−1.21	112.72
4 (9 points)	0.096	1.93	0.92	23.86
4 (8 points)	0.094	0.66	0.97	24.47

was 36 mJ/cm^2. Sommer et al. (1998, 1999) reported the dose of 20–22 mJ/cm^2 for a one log reduction of this strain. Collimated beam procedure was used in both studies. The measured D_{10} dose using annular UV reactor in food model system was comparable with earlier reported data.

5.2.4 Conclusion

In the next step, *B. subtilis* spores were used as an indicator microorganism in the biodosimetry studies using a specially designed laboratory-scale annular singular lamp UV reactor. A dose–response study in the sucrose syrup was conducted with a range of flow rates chosen to simulate the laminar flow regime observed in the pilot-scale multiple-lamp UV reactor. The collected data allowed the more accurate calculation of the UV sensitivity of *B. subtilis* spores expressed as a D_{10} and to verify the minimum UV dose delivered in the pilot-scale UV unit tested in the phase 1 of the project. The value of the UV dose of 314.8 mJ/cm^2 was lower than the value obtained in the pilot-scale reactor and accounted for the differences in mixing behavior in the laboratory- and pilot-scale reactors due to the chamber configurations. The variations of 12–15% in the absorption coefficient of the sucrose from different lots were found and need to be taken into consideration since the absorption coefficient is a factor that can affect the delivery of UV dose and thus efficacy of UV inactivation. Series-event model was more accurate than first-order model in predicting inactivation rate constant *of B. subtilis* spores.

5.3 UV inactivation of pathogenic and spoilage organisms in liquid sucrose

The sensitivity of pathogenic and spoilage microorganisms in sucrose syrup to UV light was studied and the UV dose requirements to achieve various inactivation levels of food pathogens, such as *Escherichia coli*, *Salmonella*, and *Listeria monocytogenes*, in sucrose syrup were determined.

The objective of these tests was to conduct inactivation kinetic studies with five target pathogenic organisms, including *Salmonella*, *L. monocytogenes*, nonpathogenic *E. coli*, and spoilage microorganisms, such as yeast and mold in the singular lamp reactor. Also, from the data generated, the goal was to estimate the inactivation level of target microorganisms achieved by the UV dose delivered in the pilot-scale reactor. The resistance to UV light in sugar syrup of pathogenic and spoilage microorganisms, including *Salmonella*, *L. monocytogenes*, nonpathogenic *E. coli*, yeast, and molds, was also tested.

5.3.1 Methodology

5.3.1.1 Test organisms

A cocktail of strains was used for each test microorganism (Table 5.23). Selection of strains included outbreak strains, isolates recovered from dairy products or dairy environmental samples, and strains that have been used in previous UV inactivation studies.

Table 5.23 Microbial strains of food pathogens and spoilage organisms.

Test organism	Serotype	Strain	Source
Listeria monocytogenes		ATCC 51414	Raw milk associated with listeriosis outbreak
		4-67-1	In-line ice cream
		5-115-1	Raw milk
		6-11-3A	Environmental ice cream
		8-040-4	Yogurt
Salmonella	Newport	ATCC 6962	Food poisoning fatality, United Kingdom
	Dublin	ATCC 15480	
	Montevideo	ATCC 8387	
	Typhimurium	ATCC 13311	Human feces/food poisoning
	Oranienburg	B2695	1998 ice-cream outbreak
Escherichia coli		ATCC 11229	
		ATCC 15490	
		ATCC 15922	
	K12	ATCC 25253	Previous UV inactivation work in juice (3)
		ATCC 23716	
Yeast		8a	General Mills
		9a	General Mills
		11a	General Mills
		13a	General Mills
Mold		BW	General Mills
		17-1-4	General Mills
		17-2	General Mills

5.3.1.2 Preliminary experiments to determine stability of pathogens in sucrose syrup

Preliminary experiments were conducted in the laboratory to determine if test microorganisms (*L. monocytogenes*, *Salmonella*, *E. coli*) could be successfully recovered from artificially inoculated sucrose syrup using standard methods. For each test run, 100 μL of an overnight culture of the test microorganism "cocktail" was added to c. 500 mL of sucrose syrup to achieve an initial target inoculum level of c. 10^5 CFU/mL of sucrose syrup. Inoculated sucrose syrup was mixed for 5 min by mechanical stirring, to mimic pilot plant conditions. After mixing, inoculated sucrose syrup was held at ambient temperature for a period of 1 h, to replicate worst-case conditions from time of inoculation to actual laboratory analysis during the experiments conducted with the laboratory-scale annular UV unit. Dilutions of the inoculated syrup were prepared in 90-mL sucrose syrup and selected dilutions were spread plated on TSA. Plates were incubated at 35°C for 48 h.

5.3.1.3 Preliminary experiments to determine distribution of mold spores in sucrose syrup

As the mold cultures were in a dry "powder" state, preliminary work was conducted to determine the best method for incorporating the mold spores into the sucrose syrup to ensure even distribution of spores throughout the sucrose syrup. A 0.1-g portion of each of the three mold stock cultures was aseptically weighed into a sterile centrifuge tube containing 9 mL of PW. The diluted spore suspension was mixed for 5 min using a vortex, then added to c. 2500 mL of sucrose syrup, and mixed by stirring for an additional 5 min. Serial dilutions of the inoculated syrup were prepared and selected dilutions were spread plated on acidified potato dextrose agar and incubated at RT (20°C–25°C) for 5 days.

5.3.1.4 Pathogens

Stock cultures of *L. monocytogenes*, *Salmonella* and *E. coli* stored at 4°C on TSA slants were used to inoculate 9-mL sterile TSB and incubated at 35°C for 18–20 h. This was done separately for each of the five strains in the cocktail.

After incubation, the five strains of each test microorganism were pooled by aseptically transferring 5 mL of each overnight culture to a single sterile centrifuge tube. This "cocktail" was centrifuged for 5 min at 5000 rpm, the supernatant was discarded, and the pellet was resuspended in 10-mL sterile 0.1% PW, resulting in an inoculum level of 10^9 CFU/mL. This entire volume was used for inoculation of the sucrose syrup.

5.3.1.5 *Yeast and molds*

Stock cultures of yeast stored at 4°C on agar slants were used to inoculate 9-mL TSB and incubated at RT, 20°C–25°C for 5 days. Following incubation, 40 mL of each culture was centrifuged at 5000 rpm for 5 min, the supernatant was discarded, and the pellet was resuspended in 10 mL of PW. This entire volume was used for inoculation of the sucrose syrup.

A 4-g portion of each mold spore powder was weighed into a sterile Falcon tube to which 20 mL of PW was added. The mold spore suspension was mixed for 5 min using a vortex and used for the inoculation of sucrose syrup.

5.3.1.6 *Recovery of test microorganisms from sucrose syrup (preliminary studies)*

Test microorganisms, including *E. coli*, *Salmonella*, and *L. monocytogenes*, were recovered from sucrose syrup after holding for 1 h but actual levels recovered were approximately $1\log_{10}$ lower than expected. A second experiment was conducted with *Salmonella*, comparing spread and pour plate methods of recovery, and to determine if plating samples immediately after inoculation would provide better recovery than holding in the syrup for 1 h. Results for both plating methods were comparable. Samples plated immediately after inoculation resulted in better recovery than samples held in the syrup for 1 h before plating.

5.3.2 Results

In the bench-scale studies, inoculated syrup containing the target pathogen was exposed to UV light under controlled flow conditions. In more details, to determine the sensitivity of selected food pathogenic and spoilage microorganisms to UV light microbial inoculation studies were conducted in sucrose syrup in the annular reactor at the maximum flow rate of 140 rpm or 308.7 mL/s and the minimum flow rate of 85 rpm or 151.2 mL/s. All tests' runs were conducted in triplicate and summary of results is given in Table 5.24.

The UV dose of 183.4 mJ/cm^2 delivered in the reactor at a flow rate of 4.9 gpm, or 308.7 mL/s, resulted in greater than a $5\log_{10}$ reduction of each tested pathogen. The estimate of the dose required for $1\log_{10}$ reduction of the pathogenic microorganisms showed that in pilot-scale

Table 5.24 UV inactivation of food pathogenic and spoilage microorganism in sucrose

Microorganism	Flow rate (rpm)	UV dose (mJ/cm^2)	Volume average log_{10} reduction				Mean log_{10}
			1	2	3	4	
Escherichia coli	140	183.5	>5.58	>5.57	>5.54		>5.56
	85	374.7	>5.74				>5.74
Salmonella	140	183.5	>5.68	>5.65	>5.43		>5.59
Listeria monocytogenes	140	183.5	>5.77	>6.18	>6.19		>6.05
Yeast	140	183.5	3.51	3.71	3.56	3.42	3.55
	100	256.9	3.38	3.23	3.33	3.98	3.48
Mold	140	183.5	2.48	2.63	2.01		2.37
	100	256.9		2.79	2.89		2.84

commercial UV reactor, at the maximum flow rate, greater than $10\log_{10}$ reduction of *E. coli*, *Salmonella* sp. and *L. monocytogenes* can be obtained.

Spoilage yeasts and molds were more resistant to UV light treatment and molds being the most resistant to UV treatment. In average, 3.5log and 2.5log of reduction of yeasts and molds respectively were achieved at the tested UV doses. Based on similar studies conducted with the pathogenic bacteria, it was determined that approximately a $4\log_{10}$ reduction of the spoilage microorganisms tested can be achieved in the pilot-scale reactor at the minimum delivered UV dose.

The obtained data of the inactivation of pathogenic and spoilage microorganisms at the flow rate of 308.7 mL/s allowed to estimate and compare the sensitivity of the tested microorganisms to UV light in sugar syrup. The comparison of UV sensitivity of the tested microorganisms based on the estimation of UV dose needed for $1\log_{10}$ reduction is shown in Fig. 5.16.

It can be seen that molds were the most resistant microorganism to the UV light among tested microorganisms. Their UV resistance was somewhat close to the resistance of *B. subtilis* spores. The tested cocktail strains of *E. coli*, *Salmonella*, and *L. monocytogenes* had similar resistance in liquid sucrose (approximately 30 mJ/cm^2) and were the most sensitive to the applied dose of UV light.

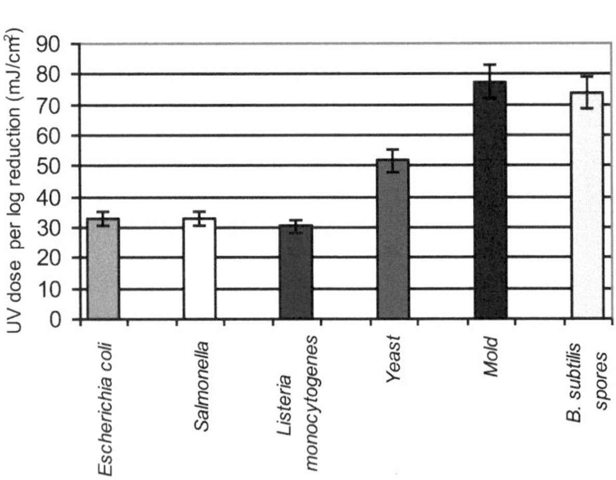

Figure 5.16 Estimated sensitivity of tested microorganisms to UV light in sucrose syrup expressed as UV dose per $\log_{10}$ reduction at the flow rate of 308.7 mL/s.

5.3.3 Conclusions

Knowledge of the value UV dose in multiple-lamp UV system and estimate of the UV dose for $1\log_{10}$ reduction of the tested pathogenic and spoilage microorganisms enabled the determination of the expected microbial reduction after 1 pass through the pilot-scale commercial UV unit at the maximum flow rate of 35.7 gpm, when a minimum UV dose of 314.8 mJ/cm^2 was delivered in that unit (most conservative estimate). The estimated reduction is shown in Fig. 5.17; at the maximum flow rate, higher than $10\log_{10}$ reduction of pathogenic microorganisms would be achieved in the pilot-scale multiple-lamp UV reactor. In terms of spoilage microorganisms, a $4\log_{10}$ inactivation of molds and $6\log_{10}$ inactivation of yeasts can be expected at the minimum processing conditions.

Also, it was found that *E. coli*, *Salmonella*, and *L. monocytogenes* were sensitive to UV light in sucrose syrup whereas molds were the most resistant to UV among tested microorganisms. The estimation of UV fluence and microbial inactivation in the pilot-scale UV reactor showed that UV light treatment is a feasible nonthermal method of control of pathogenic and spoilage microflora typical for liquid sucrose.

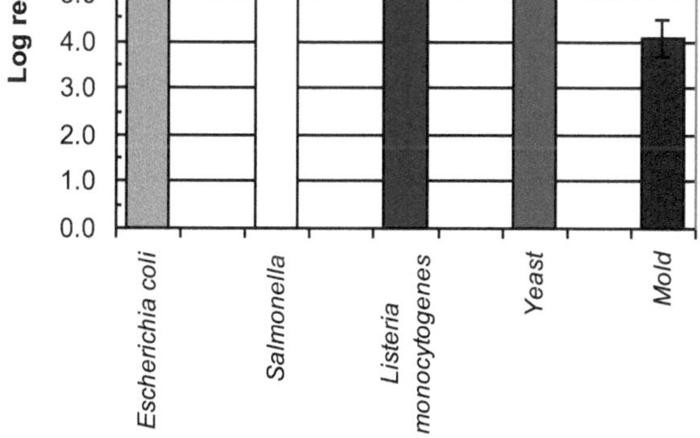

Figure 5.17 Estimated UV inactivation ($\log_{10}$ reduction per pass) of pathogenic and spoilage microorganisms in the pilot-scale multiple-lamp UV reactor.

References

American Type Culture Collection, 2005. Product Information Sheet for ATCC® 6633. Manassas, VA.

Bird, R.B., Lightfoot, E.N., Stewart, W.E., 2002. Transport Phenomena. John Wiley & Sons Inc, New York.

Chang, J.C.H., Ossoff, S.F., Lobe, D.C., Dorfman, M.H., Dumais, C.M., Qualls, R.G., et al., 1985. UV inactivation of pathogenic and indicator microorganisms. Appl. Environ. Microbiol. 49, 1361–1365.

Claus, D., Berkeley, R.C.W., 1986. Genus *Bacillus*. In: Sneath, P.H.A. (Ed.), Bergey's Manual of Systematic Bacteriology, Vol. 2. Williams & Wilkins, Baltimore, MD, pp. 1120–1125.

Hamouda, T., Shih, A.Y., Baker Jr., J.R., 2002. A rapid staining technique for the detection of the initiation of germination of bacterial spores. Lett. Appl. Microbiol. 34, 86–90.

Harm, W., 1980. Biological Effects of Ultraviolet Radiation. Cambridge University Press.

Nicholson, W.L., Galeano, B., 2003. UV resistance of *Bacillus anthracis* spores revisted: validation of *Bacillus subtilis* spores as UV surrogates for spores of *B. anthracis* Sterne. Appl. Environ. Microbiol. 69, 1327–1330.

Severin, B.F., Suidan, M.T., Engelbrecht, R.S., 1983. Kinetic modeling of U.V. disinfection of water. Water Res. 17 (11), 1669–1678.

Severin, B.F., Suidan, M.T., Rittmann, B.E., Engelbrecht, R.S., 1984. Inactivation kinetics in a flow-through UV reactor. J. WPCF 56 (2), 164–169.

Sommer, R., Cabaj, A., 1993. Evaluation of the efficiency of a UV plant for drinking water disinfection. Water Sci. Technol. 27, 357–362.

Sommer, R., Cabaj, A., Schoenen, D., Gebel, J., Kolch, A., Havelaar, A.H., et al., 1995. Comparison of three laboratory devices for UV inactivation of microorganisms. Water Sci. Technol. 31, 147–156.

Sommer, R., Haider, T., Cabaj, A., Pribil, W., Lhotsky, M., 1998. Time dose reciprocity in UV disinfection of water. Water Sci. Technol. 38, 145–150.

Sommer, R., Cabaj, A., Sandu, T., Lhotsky, M., 1999. Measurement of UV radiation using suspensions of microorganisms. J. Photochem. Photobiol. 53 (1–3), 1–5.

CHAPTER 6

Overview of other novel processes and validation approaches

Tatiana Koutchma
Novel Food Processing Technologies, Agriculture and Agri-Food Canada, Guelph, ON, Canada

6.1 Introduction

Other technologies being investigated for commercial use as sterilization or pasteurization techniques are often derived from technologies that have existed for some time but have been resurrected because of new developments and higher food quality and safety requirements.

As with many of the previously discussed methods, the refinements and developments for food applications were originally developed for other industries and have been subsequently adapted to food-processing uses. Some of these methods include irradiation, advanced microwave (MW) and radiofrequency (RF) heating, pulsed electric field and electric discharge, ultrasound, cold atmospheric plasma (CAP), and multiple-hurdle processing methods that may combine several of these methods and often include conventional thermal heating or chemical additives. Most of these new processing technologies are being tested on an experimental basis for use with fluids, semifluid food, and solid and powder products.

This chapter focuses on an overview and brief discussion of the principles, commercial applications, and main validation approaches of two commercially available technologies, irradiation and MW heating and emerging treatment using CAP. The pros and cons of each technology are presented, along with an update on the existing regulatory approvals and validation procedures.

6.2 Radiation, regulations, and validation

Food irradiation has been proven to be a safe and effective nonthermal process for enhancing food safety and extending the shelf-life of a wide variety

Validation of Food Preservation Processes based on Novel Technologies
DOI: https://doi.org/10.1016/B978-0-12-815888-3.00008-7

of foods. Other terms commonly used for irradiation are "cold pasteurization" and "irradiation pasteurization or sterilization." The term "electronic pasteurization" has been coined for e-beams and X-ray technologies with the added benefit that the electronic source stops radiating when switched off. Irradiation is also often referred to as "ionizing radiation" because it produces rays or particles strong enough to dislodge electrons from atoms and molecules, thereby converting them into electrically charged ions.

Gamma rays from radioactive nuclides, energetic electrons from particle accelerators, and X-rays emitted by high-energy electrons are suitable kinds of radiant energy because these three ionizing energy sources can produce similar effects in any irradiated material. High-energy electrons, gamma, and X-rays can kill or inactivate bacteria, spores, fungi, and insects by breaking molecular bonds in the microbial DNA.

6.2.1 Gamma rays

Gamma irradiation technology was patented more than a century ago (in 1906) and was one of the first nonthermal technologies to be thoroughly tested, validated, and adopted by the medical and food industries over the past 60 years. Despite this, there is a lot of misunderstanding and concerns associated with the consumer acceptance, transport, storage, occupational hazards, and disposal of radiation sources. Gamma rays for food treatment are produced by radioactive isotopes of cobalt-60 or cesium-137. Gamma sources are specified in terms of their activity measured in curies (Ci); 1 MCi is a moderate-sized source. The gamma rays inactivate microorganisms in the "direct hit" method and in the "indirect hit" method. In the direct hit method, the gamma ray directly collides with the genetic material of the bacterial cell, breaking the DNA. In the indirect method, the gamma ray passes very close to the DNA, ionizing a water molecule. The reactive components of the ionized water molecule react with the DNA, resulting in lethal damage. In practice, both methods contribute to the inactivation of bacteria. In low-moisture environments, most notably bacterial spores, the indirect method has little effect, which in part accounts for the higher radiation resistance of bacterial spores.

In the United States, gamma irradiation has been widely used for decontamination of spices, seasonings, flour, shell eggs, poultry, shellfish, and more recently, lettuce and spinach. All fruit and vegetables are approved for irradiation at specified maximum dose levels to delay maturation (ripening) and for disinfestation.

6.2.2 X-rays and e-beams

The advantage of X-rays and e-beams is that they are generated from machine sources. X-ray and e-beam sources are specified by beam power: the range from 25 to 50 kW is typical for food applications. X-rays are produced by reflecting a high-energy stream of electrons off a target substance—usually one of the heavy metals—into food. X-ray irradiation is used as an alternative to methods that use radioactive materials and generated from machine sources operated at or below an energy level of 5 MeV. X-ray irradiators are scalable and have deep penetration comparable to cobalt-60. The choice of a radiation source for a particular application depends on such practical aspects as thickness and density of the material, dose uniformity ratio, minimum dose, processing rate, and economics.

E-beam irradiation uses electrons accelerated in an electric field to a velocity close to the speed of light and is generated from machine sources operated at or below an energy level of 10 MeV. Since electrons are particulate radiation, they do not penetrate the product beyond a few centimeters, depending on product density. The product is exposed to the e-beam when it moves on a conveyor. Single or double beams are used to solve issues of packaging thickness.

A major advantage of electronic irradiation is that food can be processed in its final packaging, thereby reducing, or eliminating entirely, the possibility of recontamination following this treatment. This unique operational capability makes irradiation particularly suitable for cold pasteurization of ready-to-eat (RTE) foods, such as hot dogs, and other deli items, at risk of contamination with *Listeria monocytogenes* during post-process slicing and packaging operations. Irradiation treatment is very effective in killing microbial pathogens, such as *Escherichia coli* O157:H7, *L. monocytogenes*, *Salmonella* spp., and *Vibrio* spp., among others, that are significant contributors to foodborne illness.

Food irradiation by high-energy electrons with energy upto 5–10 MeV has been proven to be a safe and effective process for increasing food safety and extending product shelf-life. Low-energy electrons (energy of 300 keV or lower with extremely small penetration capacity) can supplement food irradiation technologies with a novel sanitary and surface treatment approach. "Soft" electrons can inactivate microorganisms on the surfaces, which can be frequently contaminated with spoilage, vegetative bacteria, and heat-resistant spores. Low-dose electrons can be used as an alternative to steam or chemicals for the elimination of microorganisms on the surface of grains, dehydrated vegetables, spices, and tea leaves with little quality loss. The e-beam aseptic filling systems are emerging as alternatives to chemical sterilization of packaging by hydrogen peroxide. The pros and cons of four existing methods of irradiation are summarized in Table 6.1.

6.2.3 Safety of irradiation

A great deal of scientific evidence shows that irradiation does not cause a significant differences in the flavor, texture, or color of beef, poultry, or produce when irradiated at optimal levels (USDA, 1999; O'Bryan et al., 2008). No loss in nutrients, vitamins, proteins, lipids, and carbohydrates occurs in food irradiated upto 10 kGy. For a comparison, according to studies by the ARS USDA, a radiation dose of only 3.75 kGy inactivated all of the foodborne pathogens by a minimum of 5-log (99.999%), on all of the RTE meat types, which is sufficient to be labeled as pasteurized according to current regulatory requirements. Electronic pasteurization can offer solid and semisolid foods such as meat, poultry, and fish the same benefits that thermal pasteurization has brought to milk and other liquid products.

6.2.4 Validation and verification of food irradiation process

In radiation processing, the essential quantity and critical parameter is "radiation absorbed dose" that determines the intended effect of irradiation in foods. The absorbed dose that is sometimes referred to as the dose D, is the amount of applied energy absorbed per unit mass of irradiated matter at a location of interest. The absorbed dose is expressed in Gray (Gy). One Gy equals 1 Joule/kg of product. It has been proven that low doses (0.05–0.15 kGy) are enough for inhibition of potato sprouting, disinfection of insects and parasites of fruits, and delay of ripening in fresh fruits and vegetables. A medium absorbed dose (1.0–10 kGy) is sufficient for prevention of foodborne diseases through destruction and control of pathogens such as *Salmonella* spp., *Campylobacter jejuni*, *E. coli* O157:H7, *L. monocytogenes*, and *Staphylococcus aureus*. Higher doses (10–50 kGy) are used for decontaminating dry food ingredients, like spices and herbs. Doses from 30 to 50 kGy are applied for sterilization of foods for space and hospital diets at an industrial scale.

Table 6.1 Pros and cons of four irradiation technologies.

	Gamma	X-rays	High-dose e-beams	Low-dose e-beams
Source of irradiation	Cobalt-60 or cesium-137	X-ray machine	Machine source 5–10 MeV	300 kEV
Commercial units	Available	Available	Available	Available
Cost	High investment or toll 10 cents per pound	High to moderate	High to moderate	Moderate
Penetration depth	80–100 cm	80–100 cm	8–10 cm	Surface
Best applications	Bulk processing of large boxes or palletized product in shipping cartons	Bulk processing of large boxes or palletized product in shipping cartons	Sequential processing of primary packaged products in-line	Packaging aseptic filling shelf-life extension of fruits
Packaging	Pre- and postpackaging	Pre- and postpackaging	Pre- and postpackaging	Prepackaging
Advantage	High-dose uniformity	High-dose uniformity machine source	Lower cost machine source	Lower cost machine source specific applications
Limitations	Consumer perception labeling	Consumer perception labeling	Lower dose uniformity labeling	Surface treatment

In validation of the irradiation process the lowest and the highest absorbed doses have to be specified. The lowest absorbed dose (D_{min}) required to achieve a desired effect is termed the lowest acceptable dose. The highest absorbed dose (D_{max}) acceptable is determined based on the sensory and functional properties of the product that must not be impaired by too high a dose. Their values are based on food irradiation practical experience and are specified by food technologists for each combination of process and product (IAEA, 2002).

The absorbed dose has to be specified in the particular material of interest in specific irradiation conditions, because different materials have distinctive radiation absorption properties. For food irradiation, the material of interest is generally specified as water.

According to the IAEA, the objective of validation procedures is to establish documentary evidence that the irradiation process has achieved the desired results. The key element of validation practices is a well characterized reliable dosimetry system that is traceable to known national and international dosimetry standards. Only such dosimetry systems can be used to establish the required documentary evidence. The purpose of dosimetry is to ensure that the absorbed dose requirements for a particular product and process can be satisfied. The absorbed dose received by every part of the product must be between the two specified dose limits, D_{max} (limit) and D_{min} (limit). The ratio of the maximum dose and the minimum dose is also calculated.

Due to the nature of the commercial irradiation process, fluctuations in the values of process parameters are expected and will affect the dose distribution. The distribution of absorbed dose in irradiated food products depends on plant design, type and kind of product, and type and energy of radiation. These factors will not normally vary during a given irradiation treatment. The recommended methods of measuring dose distributions inside the product include experimental data from the placement of thin films or thin-film strip dosimeters throughout the product and theoretical calculations.

Based on the above considerations, the validation program involves the four main elements:

1. Product qualification (setting the maximum and minimum dose limits);
2. Process qualification (dose mapping and selection of process parameters);
3. Facility qualification that includes characterization of the equipment, conveyor system, and dosimetry;
4. Documentation and certification.

The objectives of the irradiation process qualification are:

1. To establish process parameters (such as dwell time, electron beam parameters, conveyor speed, irradiation geometry, and product loading configuration) for the given process that satisfies the dose specifications;
2. To identify reference location(s) on (or near) the process load where a routine dosimeter can be placed for process control.

6.2.5 Irradiation facility

There are two business models for food irradiation. One is the construction of a facility to be owned and used by the processor, while the other is to use a contracted service provided by an independent service provider. Unless a very large quantity of product is to be irradiated continuously, most processors rely on contracted services. A typical irradiation facility consists of the source of irradiation, conveyor system, safety systems, and temperature-controlled storage zones. Dosimetry is the component of a total quality assurance program for an irradiation facility. The operator of an irradiation facility must show that the entire radiation process is continuously under control, and the necessary validation documentation can be provided to demonstrate that all part of the product volumes in the process load have received the specified treatment.

6.2.6 International regulations and standards

Almost 60 countries have approved or allowed food irradiation, including the United States, Canada, Australia, some European Union members, Russia, and countries in Africa and Asia. However, according to the latest statistics from the Food Irradiation Update, China leads the world in the amount of irradiated foods. China irradiates a million tons of food, and 40% of its chicken products. Pet food comes next, along with dehydrated vegetables, spices, herbs and seasonings, and seafood.

Foods that are permitted to be irradiated differ greatly between countries. In most countries, approvals are issued for foods and food categories for a specified intended processing effect on a case-by-case basis. Currently, in the United States, Canada, Australia, and New Zealand, new approvals can only be obtained in response to a petition in order to amend the existing regulation. Any relevant agency can submit a petition, but often this has is done by a food industry. Regulations regarding dose limitations for food irradiation vary greatly across countries. As an example, the United States permits 30 kGy maximum absorbed dose for spices

(Eustice, 2018). In Australia and New Zealand, minimum and maximum absorbed doses for permitted spices vary between 2 and 30 kGy, whereas in India doses from 6.0 to 14.0 kGy are allowed for elimination of pathogens. In Canada, the EU, and China, the maximum allowed dose is 10 kGy.

Also, food processed by ionizing irradiation is subject to all relevant standards and codes, including ISO, that are developed to establish uniform specifications, procedures, or technical criteria. International Organization for Standardization (ISO) is the world's largest developer of voluntary international standards. The ISO standards identified the essential practices to be implemented to achieve effective radiation processing of food products in a manner that maintains quality and yields food products that are safe and suitable for consumption. The ISO standards cover the standard practice for dosimetry in gamma, electron beam, and X-ray irradiation facilities for food processing (ISO/ASTM 51204 and ISO/ASTM 51431), and selection and calibration of dosimetry systems for radiation processing (ISO/ASTM 51261).

A new standard ISO 14470:2011, Food irradiation—Requirements for the development, validation and routine control of the process of irradiation using ionizing radiation for the treatment of food, not only provides requirements, but also guidance for meeting them. It is available from ISO national member institutes or through the ISO store (http://www.iso.org/iso/home/store/catalogue_tc/catalogue_detail.htm?csnumber = 44074). The main objectives of ISO 14470:2011 are to:

1. Provide requirements for the irradiation of food consistent with current standards and practices;
2. Provide directions for a technical agreement between the customer and the irradiator operator;
3. Establish a documentation system to support the controls on the food irradiation process.

A new standard defines the requirements for irradiation facilities, and the safety of the irradiation process and food packaging. In regards to irradiation facilities, the design shall be specified under the guidelines and in accordance with irradiation process specifications and regulatory requirements. Necessary measures shall be taken in order to prevent contamination and cross-contamination and to avoid direct and indirect contact of the food with potential sources of contamination.

According to ISO 14470, food processors should include irradiation in the food safety process whenever appropriate. The standard calls food irradiation a critical control point of an HACCP program, "contributing to the minimization of risk from the transmission of pathogenic microorganisms to consumers."

Aiming to ensure food quality and safety, a series of packaging and processing requirements, designed to manage food irradiation processes, have been outlined. For example, the standard says that "the food shall be packaged in suitable materials for each type in the irradiation process." In regards to the documentation system for the food irradiation process, the standard requires that the product being irradiated should be defined, and any changes to the product as a result of the irradiation process need to be documented so that an assessment can be made on the process that caused the alteration.

The new standard provides minimum requirements necessary to control the food irradiation process. ISO said that this new standard makes irradiation safer and would benefit manufacturers, irradiation operators, regulators, customers and consumers.

In conclusion, it should be emphasized that food irradiation is a powerful technology to prevent foodborne illnesses. Removal of labeling requirements, use of the term "pasteurization," and regulatory approvals will lead to the wider acceptance of this technology. Education and skilled marketing efforts are needed to remedy this lack of awareness of the effectiveness, safety, and functional benefits that irradiation can bring to foods. According to the consensus white paper published by Global Harmonization Initiative (Koutchma at el., 2018), one of the major challenges facing the global food sector in the use of irradiation is harmonization of regulations and equivalence of standards, doses, and labeling. Legislation needs to be harmonized and updated continuously to facilitate the effectiveness of global food chains and trade, and respond to global food safety challenges.

6.3 Advanced radiative and electromagnetic heating technologies

In conventional thermal-processing systems, heat energy is transferred through conduction and convection from a hot medium to a cooler product that may result in large temperature gradients. Heat exchangers typically utilize pressurized steam from petroleum-fired boilers with <25%–30% of the energy conversion. Five innovative heating techniques, including infrared (IR), MW, dielectric or RF, ohmic (OH), and magnetic induction (MI) heating utilize electromagnetic energy and can heat foods faster and more efficiently. Except for IR radiation and MI heating, heat is generated within the product as a result of the transfer of electromagnetic energy directly into the product. This initiates volumetric heating due to a frictional interaction between water molecules and charged ions. These methods offer a considerable speed advantage, particularly in solid foods, with high efficiency of energy conversion ranging from 60% upto almost 100%.

6.3.1 Infrared

IR radiation encompasses the portion of the electromagnetic spectrum bordering on visible light and MWs. IR can be classified into three regions, namely, near, mid-, and far-infrared (FIR), corresponding to the spectral ranges of 0.75–1.4, 1.4–3, and 3–1000 μm, respectively. The amount of the IR radiation on any surface has a spectral dependence because energy coming out of an emitter is composed of different wavelengths and the fraction of the radiation in each band, dependent upon the temperature and emissivity of the emitter.

As food is exposed to IR radiation, it is absorbed, reflected, or scattered. In general, the food substances absorb FIR energy most efficiently through the mechanism of changes in the molecular vibrational state. Water and organic compounds, such as proteins and starches, which are the main components of food, absorb FIR energy at wavelengths greater than 2.5 μm.

Because IR's penetrating powers are limited, it can be considered as a surface treatment in both liquid and solid foods. Recently, IR radiation has been widely applied to various operations in the food industry, such as dehydration, frying, and pasteurization, as well as in domestic applications, such as grilling and baking. Electrical IR heaters are popular because of their installation controllability, their ability to produce a prompt heating rate, and being a cleaner form of heat. They also provide flexibility in producing the desired wavelength for a particular application. In general, the operating efficiency of an electric IR heater ranges from 40% to 70%, while that of gas-fired IR heaters ranges from 30% to 50%. The efficacy of IR heating depends on the IR power level, peak wavelength, and bandwidth of IR heating source; and the type, depth, moisture content, and temperature of the food sample.

6.3.2 Ohmic

OH heating relies on direct OH conduction losses in a medium. The internal heating of food occurs when an alternating electrical current is passed through due to its electrical resistance. OH requires direct contact between the electrodes and the food product to be heated. The major benefits of OH technology claims include reducing heating time by 90%, uniform heating of liquids and liquids with particles with faster heating rates, reduced problems of surface fouling, no residual heat transfer after the current is shut off, low maintenance costs (no moving parts), and high-energy conversion efficiencies of upto 97%–100%. OH technology is suitable for use in applications such as the processing of a low-acid particulate product in a can, meat cooking, stabilization of baby foods, and pasteurization of milk. Commercial OH heating systems are now available from a number of suppliers. Because food components may have different electrical conductivity and thermal properties, for example, oil droplets in emulsions or fat, these components can be heated at different rates and result in nonuniform temperature profiles. In order to minimize the heterogeneity of the heating, product formulations and pretreatments should be considered. In addition, to the thermal effect, OH heating can cause electroporation of bacterial cell membranes, and some studies have shown that lower temperatures are required for the same inactivation effect compared to conventional heat treatments.

6.3.3 Magnetic induction

MI heats metal heat exchanger structures within the product flow stream. Inductively generated heat is transferred passively to the food and, unlike MW or OH heating, is not produced in the food itself. Therefore the FDA considers MI a conventional thermal process. Since MI generates heat within metals, elaborate, even flow-driven rotary heat exchange surfaces become possible. Surfaces can be patterned to process foods of any viscosity or particulate composition. MI heat exchangers made of passive metal (such as 316 stainless), provide cheap, low-heat-per-unit-surface-area heat exchangers that potentially eliminate bake-on from high-protein/high-carbohydrate foods.

The induction coil and the heat exchanger do not have direct contact. The absence of direct utility contacts allows the MI heat exchanger to slip in and out of the induction coil with simple detachment of sanitary clamps. Therefore swapping out heat exchangers for different food types or for cleaning becomes simple. MI can also heat makeup and cleaning water completely, eliminating the need for a boiler. MI heating can replace any industrial process currently serviced by steam. MI heating is all electric and, because of this, MI allows zero carbon footprint food processing, if desired.

6.3.4 Microwave

MW frequency waves are generated through a magnetron applicator at frequencies between 300 and 300 GHz and, essentially, the interaction with the food material causes the food to heat itself. Commercially available MW frequency heating includes 40.68, 433, 915, 2450, and 5800 MHz.

Domestic MW ovens commonly used for preparing RTE products operate at 2450 MHz, while most MW industrial heating processes work at 915 MHz. Simply placing a processed sample in a MW heating system and expecting it to be heated efficiently is rarely fruitful due to the complexity of the interaction of electromagnetic energy with food matrix. Nonuniformity of heating, difficulty in tracking cold spots, and unpredictable energy coupling are major issues when using MW technology.

However, rapid internal heating leaves the possibility of selective heating of materials through differential absorption and self-limiting reactions. This presents opportunities and benefits not available from conventional heating and provides an alternative MW technology for a wide variety of products and processes. This includes defrosting, drying, blanching, pasteurization, and sterilization. In addition, MW processing systems can be upto 65%–70% energy efficient. Heat can be instantly turned on and off—this constitutes important preconditions for controllability of operation along with improvements in quality over conventionally processed products.

The future of MW processing of foods appears to be strongest for special applications and it will probably be of limited usefulness as a general method of producing process heat.

6.3.5 Radiofrequency

The RF band of the electromagnetic spectrum covers a broad range of high frequencies, typically in the MHz range (1 MHz $< f \leq$ 300 MHz). The permitted frequencies for industrial applications are 13.56, 27.12, and 40.68 MHz (RF). During RF heating, the product to be heated forms a "dielectric" between two metal capacitor plates, which are alternatively charged positively and negatively by a high-frequency, alternating electric field. Processing parameters of RF heating are output power, RF, power absorption/coupling, heat capacity, dielectric properties of foods that depend mainly on frequency, temperature, water content, chemical composition, geometry, and shape of food, and applicator.

Due to the greater penetration depth, RF heating is suitable for large-size products and food trays. It was demonstrated that RF energy has adequate penetrations to inactivate heat-resistant bacteria spores in food prepackaged in 6-lbs capacity polymeric trays. As a unique benefit of RF sterilization, the total time of RF processing in 6-lbs capacity trays can be reduced to one-third the time required in conventional canning processes to achieve approximately the same level of thermal sterility for bacterial spores. RF is also known for high-energy efficiency between 50% and 60%. Promising applications of RF or dielectric heating in the food industry include blanching vegetables, thawing frozen foods, post-baking snack foods, meat processing, and pasteurizing and sterilizing liquid foods and prepacked foods. Additionally, RF heating is showing value for insect control in walnut processing.

Research groups from Washington State University, University of Georgia (United States), and Seoul National University (Korea), have been investigating the potential of RF for pathogen control in bulk low-moisture foods and ingredients. Because of the low thermal diffusivity of low-moisture foods, conventional heating methods for bulk heating of low-moisture foods are slow and inefficient. Also, low water activity makes bacterial pathogens, for example, *Salmonella*, extremely heat tolerant. Using a bench top-scale RF system for inactivation tests in wheat flour (water activity 0.25, 0.45, and 0.65) it was found that approximately 5−7-log *Salmonella* reduction can be achieved after ~8.5−9 minutes of RF treatment. Furthermore, a 39-minutes RF treatment (6 kW, 27 MHz) was sufficient to inactivate *Salmonella* surrogates in 3-kg wheat-flour samples indicating that RF heating can be effective in inactivating pathogens in low-moisture conditions.

Implementing a similar process, RF Co worked closely with the Almond Board of California to successfully develop a Macrowave™ Pasteurization System capable of providing a 5 log reduction of *Salmonella* in almond nuts and other finished products and ingredients such as various types of flour, spices, xanthan gum, protein products, as well as pasteurization systems for animal feed products.

RF heating has been broadly researched for microorganism inactivation in various food commodities such as peaches/nectarines/stone fruits, ground beef, peanut butter cracker sandwiches, and packaged vegetables. Application of RF energy enhanced quality and improved process efficiency when used for liquid product pasteurization of cow and soy milk, and orange and apple juice.

Sairem of France and Stalam of Italy offer RF defrosting systems for thawing or tempering seafood, meat, and vegetables. Tempering raises the temperature of frozen blocks of fish products to just below the freezing point to facilitate cutting and further product handling. Their systems operate at a frequency of 27.12 MHz and output power from 10 to 105 kW. RF tempering reduced the time taken by traditional methods from 10 hours to 30 minutes, improved the homogeneity of product composition and thawing uniformity, showed a significant reduction of drip loss and micronutrient loss in RF-defrosted meat compared to air- or water-thawed product, and reduced food waste and water use along with high-quality products for the consumer markets.

Despite RF technology having been available in the food industry for many years, its commercialization has been slow. Recently, because of the unique benefits and high heating efficiency of RF this thermal-processing technology has been extending its applications. Also, computer simulation has been an effective tool to study and better control electric field distribution and heating uniformity in complex food composition and advance the design of RF systems for specific applications.

RF heating is a thermal process caused by a nonionizing electromagnetic form of energy, it is not viewed as a food additive. Therefore RF-treated products can carry the certified "organic" and "natural" label and do not require regulatory approval. The RF pasteurization system can be implemented as a preventative control measure to provide a required kill step and produce better quality foods.

The differences between temperature distributions, cold and hot spot locations inside the food for IR, MW, RF, and electric heating are shown in Fig. 6.1.

The pros and cons of the radiative and electromagnetic heating are summarized in Table 6.2.

In summary, electromagnetic heating technologies offer local, all-electrical heating solutions, with zero carbon footprint, and are highly efficient processing alternatives to steam with superior process control and optimal food quality. Despite these advantages, and the fact that electromagnetic techniques have been available for many years, commercialization has been slow due to difficulties in implementing the expected benefits, validation, and lack of

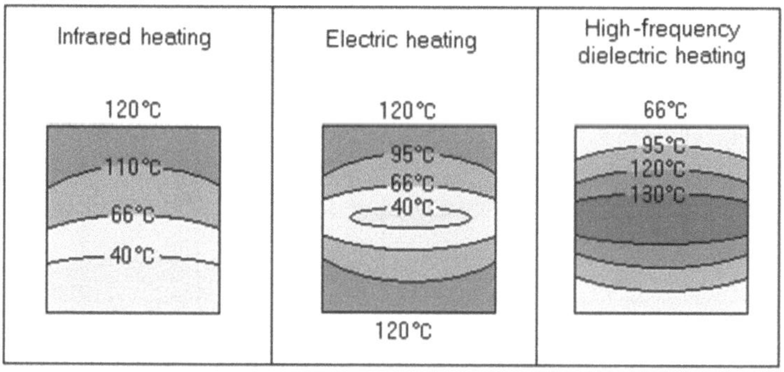

Figure 6.1 Schematics of temperature distribution upon three modes of infrared and electromagnetic heating.

information on added benefits to convince food processors. Further studies are needed to investigate temperature distributions, develop appropriate process conditions with minimal quality losses, and design an industrial system.

6.3.6 Validation principles of microwave and radiofrequency heating

6.3.6.1 Microwave process lethality

Due to the faster heating rate advantage, MW technology has been largely applied to fluid foods, such as the continuous pasteurization of milk and juices, the pasteurization of solid food products such as in-shell eggs and sterilization of low-acid prepackaged RTE meals. In general, it has been demonstrated that the destruction of microorganisms during radiative and electromagnetic heating is due to the thermal effect. The inactivation of pathogens has been demonstrated using MW technology include the following organisms: *Bacillus cereus*, *C. jejuni*, *Clostridium perfringens*, *E. coli*, *Enterococcus* spp., *L. monocytogenes*, *S. aureus*, and *Salmonella* spp. There did not appear to be any obvious "MW-resistant" foodborne pathogens. As with conventional heating, bacteria are more resistant to thermal inactivation by MW heating than yeasts and molds, and bacterial spores are more resistant than vegetative cells.

Traditional thermal destruction parameters such as decimal reduction time (D-value) and thermal resistance (z-value) are used to describe MW heating inactivation, and time and temperature parameters are critical. Traditional thermal death time tests are conducted to measure D and z-value or reported data can be used. These tests can be used for understanding attenuation or promoting effects of food matrix, pH, water activity, and chemical preservatives.

Table 6.2 Pros and cons of five advanced heating methods.

6.1 Infrared	6.2 Microwave	6.3 Radiofrequency	Ohmic	Magnetic induction
Penetration depth				
Surface	Penetration depth depends on frequency and material	Penetration depth greater than MW	Volumetric	Volumetric
Energy conversion (%)				
40–70	65–70	50–60	97–100	95
Best applications				
Baking surface pasteurization	Continuous and in-package sterilization and pasteurization viscous products	Insect control defrosting pasteurization and sterilization of packaged products	Liquids with particulates solid products	Liquids of any viscosity and with particulates
Packaging				
Prepackaging	Pre- and postpackaging	Pre- and postpackaging	Pre- and postpackaging	Prepackaging metal packaging
Advantages				
High efficiency fast commercialized	Fast heating high efficiency commercialized	Large objects	High efficiency and uniformity	Heats metal heat exchanger not the product
Limitations				
Low penetration depth	Nonuniformity	Nonuniformity low level of commercialization	Metal diffusion from electrodes Not for all products	Low level of commercialization limited research for foods

When the thermal resistance of a microorganism is known, use of a method similar to a traditional thermal process is possible to establish "MW preservation specification" and meet required specific logarithmic microbial reduction (SLR) in numbers of target microorganisms. Processing time F_p or target process value is defined by the initial load of target organisms (N_o), the end-point of the process (N_F), and the logarithmic resistance of target bacteria under defined conditions (D_T) at reference temperature, as given in Eq. 6.1.

$$F_p = D_T(\log N_o - \log N_F) = (\mathrm{SLR}) \times D_T \tag{6.1}$$

The equivalent time F or lethality necessary for thermal treatment is determined by integration of the measured time-temperature history $T(t)$ in the cold spot of the product using Eq. 6.2

$$F = \int_o^t 10 \frac{T(t) - T_R}{Z} dt \tag{6.2}$$

The validation of the processing time F_p under actual production conditions requires to accumulate a minimum lethality value under worst-case conditions (cold spot, low process temperature, etc.) of F. The measured treatment time F_p under similar worst-case conditions that is required to achieve a target SLR should meet condition (Eq. 6.3):

$$F > F_p \tag{6.3}$$

6.3.6.2 Microwave heating characteristics

The main disadvantage of MW heating is that microbial survival of various MW treatments was associated with the nonuniformity of the temperature distribution during the process. The lack of uniformity of heating has been a significant technical hurdle. Achieving heating uniformity remains a major challenge in the research and development of advanced heating technologies. This nonuniform heating results from the varying absorbing properties of foods and the surrounding air, as well as from the difference in the dielectric properties of different food constituents and their dependency on temperature. To minimize nonuniformity of MW heating the water immersion technique has been used in combination with higher penetration at a frequency of 915-MHz MWs.

Characterization of the heating parameters of the MW systems is essential to determine the most efficient processing conditions and select the optimal treatment. Heating characteristics of the MW process include spatial and time-temperature curves during transient and steady-state periods, heating rates, absorbed power, and coupling efficiency defined as a ratio of absorbed energy, and heat losses to nominal incident power. Knowledge of spatial temperature distribution is critical for the determination of a location of the least heated point or so-called cold spot in the product and further calculation of process lethality. In a case of continuous-flow process, temperature rise versus flow rate/residence time and input power need to be known. In addition, the flow regime (laminar or turbulent) and residence time distribution (RTD) need to be determined.

The desirable heating characteristics and performance of an MW system can be achieved only in a properly designed applicator. The design of an optimized MW processing system can dramatically influence the location and temperature of the coldest point, which is a critical parameter in the pasteurization or sterilization process.

6.3.6.3 Validation testing

Because of the thermal nature of MW heating, microbiological validation tests are conducted using procedures that are similar to those used for thermal-processing. Also, it has been reported that there was no microorganism with unique resistance to MW processing, suggesting that classical surrogates (vegetative cells or spores) would be appropriate for process determination and validation. Bozkurt-Cekmer and Davidson (2017) indicated a few surrogates for commercially sterilized low-acid ($pH > 4.6$) foods. Putrefactive anaerobe *Clostridium sporogenes* strain PA 3679 has been used as a nontoxigenic surrogate for proteolytic *Clostridium botulinum* in the validation of conventional heat processes. It too would be used for derivation of thermal processes for MW sterilization systems as a surrogate to produce a safe product (Guan et al., 2003; Tang, 2015). The reported D-value at 121°C of 0.6 minutes for *C. sporogenes* PA 3679 is higher than that of proteolytic *C. botulinum* (D at 121°C = 0.24 minutes) and both have a z-value close to 10°C. Thus thermal processes based on sufficient inactivation of spores of PA 3679 typically provide a considerable safety margin with respect to the destruction of spores of proteolytic *C. botulinum*.

L. monocytogenes has been commonly used as a target pathogen for designing MW pasteurization processes (Tang, 2015) because of its greater thermal resistance compared to *E. coli*, *S. enteritidis*, or *S. typhimurium*. Nonpathogenic *L. innocua* strains have been used as surrogates for *L. monocytogenes* because of their behavioral and resistance similarities.

The biggest challenge of MW validation testing is measuring a temperature history and establishing a cold spot. Temperature measurement probes, chemical or biological markers, and computer simulation are current approaches to determine the temperature profiles and cold spots in MW-heated products. Fiber optical temperature sensors are often used to avoid interference from MW fields. However, this method does not work for continuous processes in which food packages move in the system. It is impractical to connect sensors embedded in moving packages to the light sources outside the pressurized system over long distances, in particular in industrial operations. The possibility of using metallic mobile temperature sensors that are commercially available was explored for conventional continuous canning operations and reported by Tang and Liu (2015). IR thermography can be a useful tool for measuring the sample surface temperature and demonstrating the evolution of temperature change. They are a non-contact type of temperature transducer, however, the accuracy is limited.

Tang (2015) reported a development and application of standard whey protein gels and mashed potato models used with chemical marker precursors M-2 that cover the range of dielectric and thermal properties similar to those of different types of foods. The color of those model foods is pale white before thermal-processing. When heated in a package with MWs, a localized brown color develops. The color intensity depends upon the intensity of heating. Positive correlations were established between local thermal lethality F value (Eq. 6.2) and color intensity as a result of M-2 formation from MW sterilization.

Computer modeling tools are often used for the development of MW-heating-based preservation process for foods due to the complexity of the interaction of an electromagnetic field (EMF), food, and packaging. While the basic components of a MW processing system generator, applicator, and control systems are standard, the interaction of food matrix and packaging with EMF and changes in fundamental food properties during processing makes development of MW processes and validation very complex. This complexity may be dealt with using an integrated approach by a process design team including a food engineer, the MW equipment manufacturer, and an EMF specialist.

RF heating has been studied for inactivation of bacteria in a variety of foods.

Similar to MW heating, a main challenge to the commercial application of RF heating technology is in heating uniformity, regulatory approval, and commercialization.

In 2011 the US FDA approved the MW-assisted thermal sterilization (MATS) process using 915 MHz. As reported, the technology immerses packaged food in pressurized hot water while simultaneously heating it with MWs at a frequency of 915 MHz. This combination eliminates food pathogens and spoilage microorganisms in just 5–8 minutes and produces safe foods with much higher quality than conventionally processed RTEs.

Wornick Foods, a manufacturer of convenience foods and customized meal solutions, was awarded a grant to establish the MATS Research and Development Center at its facility in Ohio where consumer products companies can test the MATS process.

915 Labs has the worldwide license to manufacture and market MW-assisted sterilization and pasteurization technologies. According to 915 Labs, any food or beverage that will benefit from a lower processing temperature and a reduced processing time is ideal for MATS processing. Heat-sensitive foods such as eggs, dairy ingredients, seafood, and pastas have all been successfully processed with MATS. The reported research demonstrated that, compared to conventional thermal retorting, the shortened exposure to heat during MW processing retains more nutrients, such as omega 3, B vitamins, vitamin C, and folate.

The company currently manufactures a system that performs both sterilization (MATS) and pasteurization (MAPS) treatments. The pilot-scale system has been developed for product and process development. The smallest commercial production system processes 30 packages per minute. Larger capacity production systems with a throughput of between 50 and 225 packages per minute are available by design. Actual capacities vary depending on the nature of the product, size of package, and the desired shelf-life to be achieved. In order to realize the potential benefits of MW processes and assist food companies in commercializing MW technology and integrating it in the production process, 915 Labs offers product development, validation services, and packaging solutions.

MW continuous-flow sterilization of homogeneous and particulate-containing high- and low-acid viscous products has been developed using a frequency of 915 MHz. US FDA acceptance has been also granted and cleared the way for commercial applications for pumpable low-acid foods.

The future of MW processing of foods appears to be strongest for special applications, and it will probably be of limited usefulness as a general method of producing process heat. MW-based pasteurization and sterilization may be the technology that will enable e-commerce providers to deliver high-quality food directly to consumers through their traditional distribution channels.

6.4 Cold atmospheric plasma

CAP technology, also called nonthermal plasma (NTP) or low-temperature plasma, is a novel food-processing technology that has been mainly studied for surface treatment.

Plasma is a matter that contains partially or wholly ionized gas with a net neutral charge and is often referred to as the fourth state of matter as it shares properties similar to gases and liquids. A neutral gas can be converted to plasma by the application of energy, resulting in an increase in the kinetic energy of the electrons of constituent gas atoms. This causes a cascade of collisions in the gas, resulting in the formation of plasma products of electrons, ions, radicals, and UV radiation. Two classes of plasma, namely thermal and NTP or CAP, can be distinguished on the basis of the conditions in which they are generated. This classification of plasma is based on the relative energetic levels of electrons and heavy species of the plasma. CAP (near ambient temperatures of 30°C–60°C) is obtained at atmospheric or reduced pressures (vacuum) and requires less power. CAPs are characterized by an electron temperature much above that of the gas macroscopic temperature and consequently they do not present a local thermodynamic equilibrium. The chemical composition of CAPs of nitrogen, oxygen, and carbon dioxide gas mixtures are dominated by ions O^-, O_2^-, N^+, N_2^+ N_2, O^-, O_2^-, NO^-, CO_2^+, free radicals, and highly reactive intermediate species. If water vapor is present, highly reactive species including H_2O^-, H^+, OH^-, and HO^-_2 are formed, and also cluster ions containing H_2O. CAP can be generated by an electric discharge in a gas at lower pressure or using MWs. Typical approaches for plasma generation at atmospheric pressure include the corona discharge, dielectric barrier discharges, RF plasma, and the gliding arc discharge.

CAP processing parameters include frequency, power input to the discharge, working gas type, pressure, flow, mode of operation of a CAP system (remote, direct, and electrode contact), density of formed species, and their energy and relative humidity. Materials can be exposed to CAP by direct, indirect, or remote methods. Direct cold plasma exposure occurs when the samples are placed directly under the plasma plume discharge and all plasma agents come into contact with the sample. Indirect or remote exposure requires the sample be placed at a distance from the plasma discharge or in an adjacent chamber. In this configuration, the charged particles do not play a role since they recombine before reaching the sample, and many of the short-lived neutral reactive species also do not reach the sample.

CAP bench-scale devices are designed in a batch operation mode. The continuous operation mode looks feasible for commercial-scale units when product can be placed on a conveyor and be exposed to CAP.

CAP is a unique technology to inactivate microbes on surfaces due to the synergistic effects between three possible mechanisms of inactivation including UV photons, free radicals, and reactive species. CAP can be very effective against microbial cells and spores on surfaces with >4 log reductions. According to reports, bacteria, spores, and fungi can be inactivated relatively easily by CAP. The helium atmospheric plasmas were effective in inactivating selected *Bacillus* and *Clostridium* spores with D values (decimal reduction time) ranging from 2 to 8 minutes. Among all spores tested, *C. botulinum* type A and *C. sporogenes* were significantly more resistant to plasma inactivation than other species. The D values were comparable to other sterilization methods; however, plasma sterilization offers unique applications that require low temperature and atmospheric pressure.

Based on the unique properties and advantages of CAP, it can be used as a prepackaging step for food surface treatment and also for the packaging treatments. CAP can replace UV and pulsed light technology for treatment of packaging, food contact, and food surfaces. CAP also can be an alternative to ethylene oxide gases and treatments using gamma irradiation and high-energy (10 MeV) electron beams for dry particulate foods. Those products of low lipid and fat content (i.e., dried herbs and spices and other horticultural products) in which UV and radical exposure would have a minimum impact on oxidation or other chemical changes, may offer the best opportunities for the application of low-temperature plasma to food products.

6.4.1 Validation challenges

No commercial CAP food process has been developed so far. Pilot- or commercial-scale CAP systems will necessarily present a higher level of capital cost than the laboratory bench-scale devices that have been developed to date. How capital costs will scale for application in food processing will be influenced by the specific CAP technology used, and the intended application. Potential food applications are shelf-extension of fresh produce and fresh-cut produce, dry food particulates, shell eggs, biofilm control, packaging, prepacked RTE meats, and meat and poultry carcasses. Other available commercial examples of CAP applications include sterilization and functional modification of the surfaces of biomedical materials and devices, and surface-cleaning technology for the removal of disinfection chemicals applied to medical devices manufactured from heat-sensitive plastics.

For replacing other surface technologies, CAP requires regulatory approval and consumer acceptance. The effect of the UV and radical species of plasma on the lipids and other sensitive constituents of the foods such as vitamins C and E (which are naturally occurring in most fruits and vegetables and many foods) and quality parameters remain ambiguous. The technical challenge lies in presenting the food product in an appropriate way to the CAP to take advantage of the generation of microcidal species. The suitability of CAP for treatment of high-fat/lipid-containing and other sensitive foods (where chemical changes may be induced) is doubted. Additionally, the safety of gases in direct CAP applications needs to be evaluated.

Microbial validation studies using CAP have challenges. The process key parameters include the gas type/mixture, gas density, electron density, and temperature (T_g) that govern the plasma formed and the observed process efficacy. Also, there are difficulties with process scaling up without affecting the process uniformity. Cullen et al. (2018) emphasized that in order to achieve a method that has assured uniform plasma discharge when scaling up, there must be thought put into the changes of system volume, sample surface interactions, plasma density, sample residence time, sample-to-source distances, power and current supply, and the relative cost effectiveness of such systems. Three-dimensional modeling and simulations can assist with regard to couple fluid dynamics, thermal and electric field behavior, and food surface properties. A key area of research is the "reaction products" due to plasma species interaction with the food surface and its microflora over extended exposure periods.

Because of the significant contribution of long-lived CAP species in the observed antimicrobial effect, it became apparent that techniques which could prevent their loss to the surrounding environment would significantly reduce the processing times required.

The issue of a measurable or controllable plasma dose for food materials is another challenge and one similarly faced by the field of "plasma medicine." Given the diversity of sources and species employed in the induced effects, it is challenging to select one or a number or parameters to define a plasma treatment dose. The alternative of measuring the dose absorbed by the target which is suitable for medical applications is not compatible with food analysis given their diverse product range.

Comprehensive studies on food systems have yet to be reported for the disinfection of food and food contact surfaces including characterization of chemical and toxicological effects. Also, commercial scale up, the selected equipment and conditions, and economic feasibility of CAP need to be validated with the food product and target organisms of interest. As indicated by Cullen et al. (2018), the process is multivariate, multi-time scale, time varying, and nonlinear. The most likely process control approaches for plasma processes at an industrial level are electrical and optical measurements. Electrical techniques commonly monitor electron density, ion species, or power input by measuring a current or potential change related to the plasma discharge. Optical techniques rely on either the optical emission from the plasma or an external light source such as a lamp or laser to probe the plasma species. In many cases, optical measurements attempt to characterize a specific species since optical emission (absorption) wavelengths are unique to a given atom or molecule. This diagnostics approach does not affect the process, that is, it does not perturb plasma conditions and is very sensitive, permitting measurement of very low densities of many gas-phase species.

6.4.2 Regulations and commercialization

No approved legislation or regulatory guidance for plasma treatment of food currently exists from any of the regulatory bodies globally. It should be noted also that CAP exposure can induce a range of chemical changes to food components including the oxidation of sugars to organic acids, the modification of amino acid residues in proteins, and the peroxidation of lipids, which can result in the generation of toxic metabolites such as short-chain aldehydes. Cytotoxic and mutagenic effects have been demonstrated for long-term exposure of cell lines to complex plasma-treated protein models. It will be important to establish minimal concentrations and exposure times at which an increased rate of toxic or mutagenic effects can occur and to define safe doses dependent on the application or food target in question. The approval of comparative technologies such as ozone as a direct additive to food by the US FDA can be referenced as a potential successful regulatory approval path. Indeed, some plasma devices are currently marketed as ozone systems for food applications.

Cullen et al. (2018) pointed out a number of key recommendations for plasma adoption in food processing, that in summary include five key steps:

1. Products or product groups should be subjected to individual safety assessment depending on the intended technical effect of CAP treatment.
2. The plasma process must be described with respect to its technical parameters, such as the working gas, degree of ionization, treatment geometry, exposure time, temperature, product pH value, and system layout.
3. A profile as comprehensive as possible of the plasma-induced physical, chemical composition and biochemical/microbiological changes in the food is required.
4. It should be determined whether toxic compounds are formed as a result of plasma treatment.
5. The impact on microbiological safety profile must also be taken into account

Due to the early stage of CAP technology development for food applications, estimates of the cost of implementation are based on preliminary data. In many reports CAP is referred as a cost-effective processing method due to the low equipment and operational costs. However, the details of economic analysis remain unclear with these assessments. It is known that CAP systems that use air as the feed gas are less expensive than those that use a pure gas (Ar, He, etc.), or a defined gas mixture. Operational costs specific to CAP equipment are based on the power consumption of the CAP emitters and associated equipment (power and control electronics, monitoring, cooling, etc.), the nature and flow rate of the gas used, wear and tear on electrodes, and other factors. Bench-scale systems have widely varying power consumptions, depending on the scale of the system and the CAP technology used. For air-based systems, a minimum voltage of approximately 3 kV is required to achieve ionization.

References

Bozkurt-Cekmer, H., Davidson, P.M., 2017. Microwaves for microbial inactivation—efficiency and inactivation kinetics. In: Regier, M., Knoerzer, K., Schubert, H. (Eds.), The Microwave Processing of Foods, second ed. Woodhead Publishing, pp. 220–251. Available from: https://doi.org/10.1016/B978-0-08-100528-6.00011-5.

Cullen, P.J., Lalor, J., Scally, L., et al., 2018. Translation of plasma technology from the lab to the food industry. Plasma Process Polym 2018 (15), e1700085. Available from: https://doi.org/10.1002/ppap.201700085.
Eustice, R.F., 2018. Acceptance, Use of Food Irradiation Reached New Levels in 2017. Food Safety News.
Guan, D., Gray, P., Kang, D.-H., Tang, J., Shafer, B., Younce, F., Yang, T.C.S., 2003. Microbiological validation of microwave-circulated water combination heating technology by inoculated pack studies. J. Food Sci 68 (4), 1428–1432.
IAEA. Dosimetry for food irradiation. Vienna: International Atomic Energy Agency (IAEA), 2002. 168 p.
Koutchma, T., Keener, L. Kotilainen, 2018. Discordant international regulations of food irradiation are a public health Impediment and a barrier to glogal trade. Global Harmonization Initiative. Consensus Paper. GHI Share Sheets and Consensus Documents | Global Harmonization Initiative.
O'Bryan, C.A., Crandall, P.G., Ricke, S.C., Olson, D.G., 2008. Impact of irradiation on the safety and quality of poultry and meat products: a review. Crit Rev Food Sci Nutr. 48 (5), 442–457. Available from: https://doi.org/10.1080/10408390701425698.
Tang, J., 2015. Unlocking potentials of microwaves for food safety and quality. J. Food Sci 80 (8), E1776–E1793.
Tang, J., Liu, F. 2015. Method for recording temperature profiles in food packages during microwave heating using a metallic data tracer. US Patent No. (filed 03/22, 2011-approved 03/17/2015).
USDA, 1999. Final Rule on Meat and Poultry Irradiation. USDA.

Conclusions

Despite the differences of mechanisms of physical action and microbial inactivation, there are a few common steps and approaches that can be recommended and used for the establishment and validation of novel processes. These steps are summarized below and should be implemented when validating a new process and systems using thermal or nonthermal technologies.

Step 1

The organisms of public health concern in tested foods have to be evaluated from available reported data. Their resistance to the specific treatment agent has to be determined. Review and theoretical analysis of existing reported data have to be conducted.

The output of this analysis will allow to establish the resistance order of microorganisms of concern, estimate inactivation rate parameters and approximate lethal dose of the agent for 1-logaritmic reduction, and select the most resistant food pathogen.

Step 2

Determine composition and essential physicochemical, thermal, rheological, or optical properties of tested foods, their variation as a function of formulation and dependencies on temperature, pressure, or another critical process parameter.

This will allow to define the worst-case scenario of product's composition for testing and determine the ranges of product properties variation.

Step 3

Identify and choose the appropriate microbial reduction target for safety requirement and shelf-life, such as specific logarithmic reduction (SLR) or end point approach, should be used to determine the margins of the process. Possibility of growth of selected organisms in target products should also be explored.

Step 4

Determine strains of challenge organisms for microbial validation studies

- Cocktail of five or more strains of challenge organisms and

- Challenge indicator or so-called surrogate organisms that meet following requirements
  - High resistance to lethal agent,
  - Nearly first order inactivation kinetics over the range of parameters that will be used in the process,
  - Ease of growing and enumeration,
  - Consistent and reproducible assay results,
  - Nonpathogenicity to humans,
  - Not effected by the product formulation,
  - Culturable to high concentrations,
  - Stable over long periods of time between sampling and completion of the analysis and to allow for shipment to and from the laboratory, and
  - On-site use, and enumeration without loss of viability or change in dose—response.

Step 5

If necessary, conduct laboratory scale studies using standard system and protocols to determine inactivation parameters or decimal reduction time or dose (*D*-value) of the pathogen of concern and indicator organisms and examine the effect of product parameters or composition. This indicator or surrogate organism will need to perform in a way that allows its reduction to be correlated to that of the pathogen of concern. The surrogate should be resistant enough to the process to be measurable after intervention at lethal doses is applied.

Step 6

Develop conservative estimation of the lethal dose (H_{germ}) dose to deliver the targeted microbial reduction

$$H_{germ} = D(\log N_0 - \log N_F)$$

H_{germ} is the amount of lethal dose of the treatment that has to be delivered to challenge organisms in the inoculated product to achieve SLR. D is a lethal dose to reduce 90% of relevant pathogenic population with the highest resistance. Minimum and maximum estimates of H_{germ} have to be determined to achieve target SLR, or desired shelf-life without undesired changes in quality, nutrients, or sensory attributes.

Step 7

Determine process critical factors such as

- Flow rate/regime/viscosity
  - Residence time versus exposure time and
  - Residence time is a total time particle travels through the system. Exposure time is a part of residence time when particle "sees" the exposure;
- Process temperature;
- Operating pressure;
- Irradiation dose;
- Lethal factor intensity (light or electrical field intensity); and
- Other (concentration of chemical agent).

Step 8

Process validation in pilot/commercial scale system

The overall objective of process validation is to determine the operating conditions under which a testing system delivers the validated lethal dose.

- The validated dose must be greater than or equal to the required dose to receive log inactivation credit for a target organism.
- Validation testing also establishes the operational set points used during system operations to confirm delivery of the validated lethal dose.
- Uniformity of new treatment between food batches with respect to process parameters, such as temperature, pressure, concentration, time, X-ray, or UV doses absorbed (i.e., the processing parameters that affect the treatment dose), has to be measured and evaluated. Theoretical models can be developed and used where the effect of the process parameters on the microorganism's viability can be estimated.

Step 9

Prepare the validation report

The validation report should provide detailed documentation of all validation testing results. The report should also include all elements of the validation test plan and a summary of the commercially verified properties of the system.

Recommended content of the validation report should include

- Executive summary with sufficient information to assess inactivation credit for the target pathogen(s);

- Detailed system documentation and sensors that impact process monitoring;
- Validation test plan
  - Description of the test sequence,
  - System inlet and outlet piping with dimensions,
  - Challenge microorganism stock solution preparation and enumeration,
  - Third party oversight,
  - Product quality measurement methods, and
  - Functional test methods (head loss, power, sensor reference checks);
- Results of laboratory studies
  - Microbial suspension (volume, depth, concentration), parameters measurement, and calculations;
- Results of full-scale testing
  - Flow rate, and other process parameters and
  - Inlet and outlet concentration of challenge organism;
- Quality control and checks of microbial suspension and other instruments;
- Calculation of the validated dose, log inactivation credit, and validated operating conditions;
- Measuring of the lethal dose effects on product physicochemical parameters (Brix, pH, viscosity), quality, and sensory parameters (odors, flavor); and
- Lethal dose effects on formation any undesired chemical compounds (HMF, furan, acrylamides, products of lipid oxidation, etc.).

Step 10

Verification

Verification activities are performed to ensure that system is properly operating and established processing conditions are consistently applied; lethal dose is delivered and thus confirms the efficacy of all elements of the processing system. Confirmation activities also should include actions that verify operation of system's controls and reviews of monitoring records. The process should be tested and verified over a period of operational conditions (time, shift, cleanliness, and maintenance) so that a history of acceptable operation can be developed for the process.

Step 11

Other considerations

- Equipment should be carefully selected and calibrated to minimize uncertainty;
- Preventive maintenance program should be developed.
- Measurements of flow rate, electrical power consumption, and head loss should be traceable to an independent standard.
- Uncertainty of measurements should be less than 10%.
- Sanitation
  - Line sanitation practices include preferably internal pipe and equipment cleaning methods (cleaning in place). Such processes should circulate detergent and sanitizing solutions at the precise temperatures and time to affect total control and maximum effectiveness of the line sanitation process.
- Plant quality control, Hazard Analytical Critical Control Points (HACCP), and Good Manufacturing Practices (GMP). The food processor should be able to establish appropriate HACCP and GMP procedures to deal with potential deviations and to specify some processing adjustments to deal with a possible process/product deviation.
- Quality of ingredients, water, packaging materials, and plant processes shall be carefully monitored to ensure they meet company specifications and (inter)national standards. Quality control and quality assurance departments, preferably independent from production, are responsible for the standards and specifications and monitoring of the plant quality programs.

Step 12

Process deviations

The last step to consider is the probabilities of process deviations. Sufficient understanding of the process is needed to be able to appropriately use the correct procedures to address these deviations. Such knowledge and understanding of new process performance derives during the process development and validation steps. The suggested approaches in dealing with a deviation are: first, to segregate and hold those products that were not treated according to the established process and evaluate the produced product to make sure if they are safe from potential public health risks. Second, to adjust the processing conditions whereas the treatment is being delivered in such a way as to add additional treatment to the product so that the final product has an equivalent treatment that was specified by the initial process.

Index

Note: Page numbers followed by "*t*" refer to tables.

I

J

L

V

FSC
www.fsc.org
MIX
Papier | Fördert
gute Waldnutzung
FSC® C083411

Kolibri 360 GmbH
Ferdinand-Jühlke-Straße 7
99095 Erfurt, Deutschland
produktsicherheit@kolibri360.de